Richard Dronskowski
Chemical Bonding

Also of interest

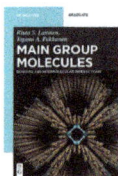

Main Group Molecules.
Bonding and Intermolecular Interactions
Laitinen, Pakkanen, 2024
ISBN 978-3-11-061327-8, e-ISBN 978-3-11-061333-9

Complementary Bonding Analysis
Grabowsky (Ed.), 2021
e-ISBN 978-3-11-066007-4

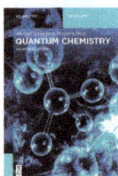

Quantum Chemistry.
An Introduction
Springborg, Zhou, 2021
ISBN 978-3-11-074219-0, e-ISBN 978-3-11-074220-6

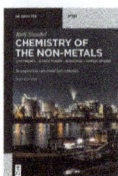

Chemistry of the Non-Metals.
Syntheses - Structures - Bonding - Applications
Steudel, Scheschkewitz, 2020
ISBN 978-3-11-057805-8, e-ISBN 978-3-11-057806-5

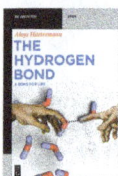

The Hydrogen Bond.
A Bond for Life
Hüttermann, 2019
ISBN 978-3-11-062794-7, e-ISBN 978-3-11-062801-2

Richard Dronskowski
Chemical Bonding

From Plane Waves via Atomic Orbitals

DE GRUYTER

Author
Prof. Dr. Richard Dronskowski
Chair of Solid-State and Quantum Chemistry
Institute of Inorganic Chemistry
RWTH Aachen University
52056 Aachen, Germany
drons@HAL9000.ac.rwth-aachen.de

ISBN 978-3-11-116653-7
e-ISBN (PDF) 978-3-11-116721-3
e-ISBN (EPUB) 978-3-11-116726-8

Library of Congress Control Number: 2023939089

Bibliographic information published by the Deutsche Nationalbibliothek
The Deutsche Nationalbibliothek lists this publication in the Deutsche Nationalbibliografie;
detailed bibliographic data are available on the Internet at http://dnb.dnb.de.

—

To my science friends in Japan,
wherever they live and work,
in the search for chemical truth

and

to the memory of Karl Jug (1939–2023)

Preface

Sometimes it is rather easy not to be completely wrong, even when alluding to the future. Almost two decades ago, when finalizing my book on *Computational Chemistry of Solid State Materials*, I closed by saying that "computational techniques are now also becoming essential tools of the synthetic experimentalist." Holy smoke, how quickly time passes by, and eventually you find yourself in a situation puzzlingly resembling what has been forecast. But that prediction was trivial, of course, anyone could have done that.

And it is true: at the present time, quantum-mechanical calculations on old and new molecules and solid-state materials are being performed everywhere and all the time, both in the academic field and in industry. There are thousands of scientists all over the globe struggling for new materials not only by experimental means but also theoretically, by sheer computation, every day. Is it tremendously difficult? That depends on the perspective. Somewhat naively, simulating such materials has become quite easy, almost routine, and the credit mostly goes to perfectly running software, many of them (such as VASP, Quantum ESPRESSO, and ABINIT) built upon the mighty combination of pseudopotentials (or, better still, PAW potentials) and plane waves, numerically efficient and accurate, too. If the structure is unknown, evolutionary and related algorithms incorporated into USPEX, CALYPSO, XtalOpt, etc. will kindly generate them. Given a powerful computer, almost every thinkable materials property can be studied, in particular *ahead* of synthesis, for compounds yet to be made. If needed, their thermodynamics at finite temperatures (and various pressures) may also be predicted given sufficiently accurate phonon calculations using PHONOPY and others. In addition to that, digital databases such as *Materials Project* or *Nomad* are being filled, permanently, automatically and accurately, such as to allow for materials genomics, if one accepts that rather funny term. Present-day researchers are well advised to look into those databases before running yet another superfluous calculation, and they should do that one more time a few weeks later because the database will then have been actualized by utilizing an even "better" (albeit more expensive) density functional. Computers have really become much faster (and some software here and there slower, thanks to Python; I am joking, of course).

And yet, something important is entirely missing, nicely illustrated using a famous quote, mistakenly attributed to Albert Einstein, dealing with everyone's knowledge and the lack of understanding. As computers and scientific software (at least some) have become more powerful and accurate, most humans feel the need to *understand* because understanding is profoundly human. Hence, the real scientific questions, despite powerful numerical theory combined with accurate measurements, are remaining essentially the same: the known polymorph α of the phase AB looks slightly

https://doi.org/10.1515/9783111167213-202

more stable than the predicted polymorph β, but *why?*[1] All our attempts to synthesize a ternary phosphide analogous to the known nitride failed, but *why?* The new carbide is too covalent, *how* can we turn it into something more ionic? *How* should we substitute the quaternary telluride such as to improve its phase-change property? And we are not going to run 5 million additional calculations, of course. I have witnessed (and tried to answer) those questions over and over again, questions posed by highly intelligent researchers not having had the privilege to sit in a course of solid-state quantum chemistry, let alone molecular quantum chemistry, not even general chemistry.

And the aforementioned challenge of understanding will become even more urgent in the coming years. Artificial intelligence has arrived, and sooner or later (probably sooner) a powerful neural network will adapt itself to the computational problem almost perfectly such that it may provide structure, energetics, and properties *much* faster than the fastest quantum-mechanical approach. The answer will be correct, at least numerically, but nobody knows *why.* And *this* problem sounds well-known, too: the late Bobby Fischer, probably one of the greatest chess players of all time, was able to rapidly analyze a chess position although his mind wasn't much faster than ours (and certainly many orders of magnitude slower than a computer) but he intuitively *understood* the game and its rules. What are the chemical rules controlling composition and structure of tomorrow's materials? What drives their chemical bondings?

As alluded to already, I have listened to many computational scientists all over the globe within the last two decades. And I feel that there is a true need to quickly explain the results of (almost) perfectly correct computational results, amazingly enough. Despite billions of integrals numerically calculated, interpreting such results is *not* done automatically but the rather dumb computer can help us to accomplish that with our own mind given a little training. In particular, we can make the computer run a handy program such as LOBSTER operating with *orbitals* to extract plenty of chemical-bonding information, and *then* we may strive for understanding by the human mind. While it also looks a little challenging to quickly train clever people – who had the chance to study quantum chemistry, let alone for solids? – to master that field, I strongly believe that one should try, and this is what I want to accomplish in this little primer of chemical bonding in solids. We need to understand, and we will understand. Dear unknown computational theorist, I do know your challenge, so let us work together; it won't take too much time. And since you are presently reading this text, we are working together already.

This little book would have been impossible without having spent a few sabbatical months in my preferred secondary homeland: Japan. My heartfelt thanks go to my professor colleagues Takahiro Yamada and Hisanori Yamane who hosted me one more time at the Institute of Multidisciplinary Research for Advanced Materials (IMRAM) at Tohoku University in Sendai. Their hospitality has been decisive in convincing me that

---

1 Answering the question by saying that α is lower in energy contains essentially no information.

returning to Japan in post-Covid times was a splendid idea, and it was truly enjoyable to do that. Given the conditions at my home institution, there is simply no time to write a book, not even a little book, because of the many everyday tasks and the cancerous growth of administrative burdens, and I am very grateful to the Rector of RWTH Aachen University (and the taxpayers of the state of NRW) to let me run away for just a few months. Likewise, I would like to thank my group at home for their patience, in particular Andreas Houben and David Schnieders, and, not to forget, Mona Marquardt for taking care of communication and administration over more than 9,000 km. Virtual meetings have made the world a lot smaller (and running away slightly more difficult, hiding is impossible these days) but a 7- or 8-h time difference helps a lot. I was working hard and busy writing this text when my dear coworkers were still asleep. And vice versa!

In addition, I would like to thank David Schnieders and Peter Müller for having performed selected calculations and, in particular, for substantial help with some figures here and there; thank you to Hicham Bourakhouadar, too. Large parts of the book are based on prior calculations and made possible only by the LOBSTER computer code, a joint effort by many, in particular, Stefan Maintz, Volker Deringer, Marc Esser, Ryky Nelson, Christina Ertural, Peter Müller, and Andrei Tchougréeff. You can take pride in having put together a highly useful academic "product" with a zero price tag.

A final heartfelt thank you goes to Hisanori Yamane, David Schnieders, Rainer Pöttgen, Eugen Schwarz, Christoph Bannwarth, Gerhard Raabe, and Alex Corkett for having perused the book and finding stupid mistakes. Likewise, I would like to thank Mona Marquardt and Linda Reitz for checking internal consistency and carefully proofreading the book one more time; if inaccuracies are still to be found, I am the one to blame. Thanks also to Kristin Berber-Nerlinger and Ria Sengbusch at De Gruyter for editorial help. And thank you to my good ol' CF-MX4 for having worked lastingly over many years.

I do hope that you enjoy reading, it has been a real pleasure to put that little primer together. And the professional theorists may forgive me for simplifying things quite a bit here and there, there was no other choice.

Long live chemistry. Long live experiment. Long live theory. Give me understanding, in addition to numbers (*almost* Coulson).

Richard Dronskowski   Aachen, May 2023

Contents

List of abbreviations

AFM	Antiferromagnetic
AIM	Atoms in molecules
AO	Atomic orbital
ASA	Atomic-spheres approximation
bcc	Body-centered cubic
BCS	Bardeen–Cooper–Schrieffer
BI	Bond index
B3LYP	Becke three-parameter Lee–Yang–Parr (functional)
BO	Bond order
CALPHAD	Calculation of phase diagrams
CBM	Conduction band minimum
CC	Coupled cluster
CI	Configuration interaction
CNDO	Complete neglect of differential overlap
CNEE	Complete neglect of essentially everything
CO	Crystal orbital
$COBI^{(n)}$	Crystal orbital bond index (over n atoms)
COHP	Crystal orbital Hamilton population
COOP	Crystal orbital overlap population
+D	Dispersion (correction)
DFT	Density functional theory
DMFT	Dynamic mean field theory
DOE	Density of energy
DOS	Density of states
EA	Electron affinity
ELF	Electron localization function
EN	Electronegativity
ESP	Electrostatic potential
ESR	Electron spin resonance
fcc	Face-centered cubic
FM	Ferromagnetic
GGA	Generalized gradient approximation (functional)
GP	Gross population
GST	Chemically weird name for Ge-Sb-Te solid solution
GW	Many-body theory involving the Green's function
hcp	Hexagonal close packed
HF	Hartree–Fock
HOMO	Highest occupied molecular orbital
HSE	Heyd–Scuseria–Ernzerhof (functional)
$ICOBI^{(n)}$	Integrated crystal-orbital bond index (over n atoms)
ICOHP	Integrated COHP
IE	Ionization energy
IpCOHP	Integrated projected COHP
IST	Chemically weird name for In-Sb-Te solid solution
LAPW	Linearized augmented plane wave
LCAO	Linear combination of atomic orbitals
LDA	Local-density approximation (functional)

https://doi.org/10.1515/9783111167213-204

LMTO	Linearized muffin-tin orbital
LOBSTER	Local-orbital basis suite toward electronic-structure reconstruction
LUMO	Lowest unoccupied molecular orbital
MBJ	Modified Becke–Johnson (functional)
MO	Molecular orbital
MPx	Møller–Plesset (perturbation of order x)
NMR	Nuclear magnetic resonance
NP	Net population
OP	Overlap population
PAW	Projector augmented wave
PBE	Perdew–Burke–Ernzerhof (functional)
PCM	Phase-change material
pCOHP	Projected COHP
pCOOP	Projected COOP
pFC	Projected force constant
RVB	Resonating valence bond
SCAN	Strongly constrained and appropriately normed (functional)
TB	Tight binding
TRIP	Transformation-induced plasticity
TWIP	Twinning-induced plasticity
U	Hubbard parameter
VBM	Valence band maximum
vdW	van der Waals
VEC	Valence-electron count
XRD	X-ray diffraction
YBCO	Chemically weird name for Y-Ba-Cu-O superconductor

1 How to read (and not to read) this book

Reading this book is trivial; one simply starts with the first page and stops at the last. But any decoding is encoding, so let me recommend the following course:

In writing this book I have imagined a highly skilled computational scientist who is perfectly capable to run sophisticated first-principles calculations targeted at structure and energetics of whatever kind of solid-state material (and also a molecule if that is needed). That being said, this person (= you) **does** know about quantum mechanics, density-functional theory, exchange-correlation functionals, plane waves, pseudopotentials, band structures, densities of states, that kind of stuff. Clearly, you do not need any instructions as regards the latter techniques, you know already. Nevertheless, in order to put us on the same level and at least agree on a common language, I have formulated a brief chapter in which I hastily summarize all essential quantum-mechanical or quantum-chemical concepts. As you will realize, you **do** know a lot already (and if you don't, no problem, we will take care of that). Some will even realize that reading Chapter 2 was not at all necessary. If that is the case, one point for you.

The subsequent Chapter 3 is slightly different in character, though. To be able to really understand the objects mentioned above, **chemical thinking** is necessary, that is, thinking associated with decomposing everything into atoms and bonds, there is no other way. **There is no other way.** Let us not forget that there is a tremendously rich and successful chemical tradition of understanding molecules, even at a time when it had not yet been understood (by our physicist friends) that molecules are actually made of **atoms**, and the atomic character of the world had not yet been proven, at least not within physics (but in chemistry already for many decades, almost forgotten these days). That being said, we will reiterate classical ways of thinking about chemical bonding and then figure out how these concepts look through the prism of quantum chemistry. There will be a few interesting surprises; this I can assure you. Ionic bonding may look trivial but only because of its one-center character (which lends itself to classical electrostatics) and the hidden necessity of Pauli repulsion preventing structural collapse. Two-center interactions show up for the case of covalent bonding, and multicenter interactions will eventually turn covalency into metallicity; it is that trivial, but there are also alternatives to metallicity, not often thought of but well-known for molecules. And then there are interactions out of nowhere (van der Waals) and hybrid interactions (H bonds and similar species). And all that makes up our world.

Eventually, the wealth of chemical bonding will be illustrated in a much larger Chapter 4 intended to illustrate and also learn how the aforementioned theoretical frameworks come together. We will witness "ordinary" and "extraordinary" materials, and it will be fun to apply various chemical quantifiers such as atomic charges, Madelung energies, crystal orbital overlap populations, crystal orbital Hamilton popu-

https://doi.org/10.1515/9783111167213-001

lations, crystal orbital bond indices, fatbands, and the density of energy to those materials, all intended to understand them better. Doing that has been made a lot simpler than 15 years ago, in part due to the existence of the LOBSTER (Local-Orbital Basis Suite Towards Electronic-Structure Reconstruction) computer program, developed and distributed **for free** by my research group at Aachen. If you haven't already, try and download LOBSTER at www.cohp.de and contribute to understanding the atomistic world.

As regards the kind of materials presented in the aforementioned large chapter, I am imagining the typical objects of solid-state chemistry, mostly "inorganic" in nature, and I am also assuming them to be crystalline (hence, no rubber, sorry for that), their structures being nicely resolved on an atomistic basis because this will make atomistic calculations much easier, that is, possible, already today. In case the wealth of solid-state chemistry is new to the reader, up-to-date literature (Dronskowski et al., 2017) is available, also in terms of crystallographic structural characterization using single crystals (Englert, 2017) and powderous matter (Dinnebier et al., 2017). And there is the rich field of inorganic structural chemistry, worth taking a look (Müller, 2006).

So, please get yourself ready and your computer running. Please check whether we speak the same language in terms of sheer calculations. If you stumble over a concept you have not been taught, chances are pretty high that I covered that in the Appendix dealing with certain "treats" of general and quantum chemistry. Here I have summed up, in an extremely condensed way, what you could have learned in those chemistry courses taught at a university, and it may help you understand what your friendly chemistry colleagues really **mean** when they talk about, say, the octet rule or a "formal" charge (unlike an oxidation number, by the way). Like in any science, there is a history of terms, and one better knows them. And then please carefully study chemical thinking and how chemical understanding can be extracted from the quantum-mechanical calculation as long as some chemical concepts are conveniently formulated. As said before, please study the seven different material classes I have suggested to look at, it will be fun. And in case you use LOBSTER to reproduce those results or even extract more chemical knowledge, I would be truly happy.

A final word intended to those who may glimpse into the electronic version of this book. Yes, it is possible to use your favorite PDF reader and then just search for keywords such as "covalency" or "ICOHP" or "fatband" and suchlike, and then jump from place to place and slap together a few broken pieces of information. This is how a few people – and I hate to say it, they are often lacking experience – proceed. **Please do not read like that, it simply doesn't make sense**. I thought for quite a while about how to introduce the reader to understand solid-state chemical bonding, and there is a reason for putting this very chapter **after** that very chapter, easy to grasp when everything – the entire picture – is visible from the least distorted perspective, and I guess I also know how to get there. Thank you so much. Now, off to work!

2 Calculating molecules and solids

Sometimes it does not hurt to repeat the truly essential things. In what follows, let me please reconsider the basics of electronic-structure theory, both for molecules and solids. This is intended to briefly recall what we know and how we usually phrase it, simply to make later understanding just a little easier. It pays off to start with molecules and then continue with periodic solids, for a number of reasons. In case you are a true theory professional, you may either choose to skip this entire session or you may be amused by how I boil down the difficult theory to a **very** elementary level, and it may look as if I were not paying too much respect to those ingenious theorists having paved the way; I do respect what has been accomplished, however, let me make that very clear, but I need to simplify. And please forgive me if you find it too trivial, on purpose I am talking to a highly intelligent non-chemist here and now.

Let us assume that we want to calculate the electronic structure of a tiny system composed of atoms (so everything is made up from nuclei and electrons), an entity usually dubbed **molecule** by the chemists, such as H_2O or NH_3 and suchlike, and we are **not** dealing with spectroscopic questions but just want to get the molecule's structure and its total energy right. In this case, the time-independent wave function $\Psi(r)$ representing the stiff molecule is our target, to be sought for by solving the stationary Schrödinger equation (2.1) (1926a; 1926b), almost a century ago:

$$H\ \Psi(r) = \left[-\frac{\hbar^2}{2m}\nabla^2 + V(r) \right] \Psi(r) = E\ \Psi(r) \tag{2.1}$$

The Hamiltonian operator H describes the **entire** energetics of the molecule, including the kinetic and potential energies of both electrons and nuclei. For H_2O, there would be 3 nuclei and 2×1 (H) + 8 (O) = 10 electrons. More pompously expressed, the Schrödinger equation corresponds to an eigenvalue problem of a linear elliptic differential operator with singular potential in dimension $3N$ (with N = number of electrons, 10 for H_2O), further complicated by spin and the Pauli principle (1925).[1] In fact, it is so difficult a problem that it makes **any** computer explode (for reasonably sized molecules). Sure, the problem may be greatly simplified by first freezing out the nuclear coordinates; for H_2O, we would fix the three nuclei in space, they are then resting still. This so-called Born–Oppenheimer approximation (Szabo & Ostlund, 1989) is a rather fine choice unless spectroscopy is important, but we are not interested in that, as said before. Then we can also set the Coulomb repulsion between the nuclei (which is a constant term) to zero; it can later be added to the solution once it is available. By

---

1 Here I have covertly ignored the nuclear degrees of freedom already; there are even more dimensions.

https://doi.org/10.1515/9783111167213-002

doing so, the Hamiltonian has become **electronic** only, the reason for dealing with what we dub electronic-structure theory:

$$H = -\sum_{i=1}^{N} \frac{\hbar^2}{2m} \nabla_i^2 - \frac{1}{4\pi\varepsilon_0} \sum_{i=1}^{N} \sum_{A}^{M} \frac{Z_A e^2}{r_{iA}} - \frac{1}{4\pi\varepsilon_0} \sum_{i=1}^{N} \sum_{j>i}^{N} \frac{e^2}{r_{ij}} \qquad (2.2)$$

Please carefully look at eq. (2.2): only the kinetic energy of the electrons (left term after the equation sign), the Coulomb potential between electrons and nuclei (middle term), as well as the repulsion between the electrons (right term) are left; for electrons we use lowercase (i, j) letters and for nuclear uppercase (A) letters. Now, the true professionals then go for **atomic units**[2] and drop a number of natural constants ($\hbar$, m, ε_0, and e) but this will not bother us any longer, to be found in books on molecular or solid-state quantum chemistry (Szabo & Ostlund, 1989; Mayer, 2003; Dronskowski, 2005). To prepare ourselves for tackling a complicated molecule, we first consider a simple **atom** and approximate the wave function $\Psi(r)$ by a so-called Hartree product of one-electron wave functions (also dubbed atomic orbitals) $\varphi_a(r_1)$ occupied by individual electrons as given in

$$\Psi(r) = \varphi_a(r_1)\varphi_b(r_2)\varphi_c(r_3)\cdots \qquad (2.3)$$

This is not a bad **ansatz** at all because it brings us close to the quasi-electrons used by the chemists (Primas & Müller-Herold, 1984), and we need to add that the aforementioned atomic orbitals are likewise solutions of the Schrödinger equation with some nuclear charge (possibly screened) but only **one** electron, the so-called one-electron problem, and these sets of atomic orbitals together with the **aufbau** principle (the atomic orbitals need to be correctly "filled" with electrons) let us understand the periodic table of the elements.[3] If our atom were the oxygen atom, those orbitals would be called the 1s, the 2s, and the three 2p orbitals, so we know them already. Now, coming back to the aforementioned **molecule**, however, one would approximate the **molecular** one-electron orbitals $\psi_i(r)$ by a **linear combination of atomic orbitals** (LCAO) more typically dubbed $\varphi_\mu(r)$:

$$\psi_i(r) = \sum_{\substack{A \\ \mu \in A}} \sum_{\mu=1}^{n} c_{\mu i} \varphi_\mu(r) \qquad (2.4)$$

The symbol μ runs over all orbitals over all atoms A. Some theorists stress, for good reasons, that the atomic orbitals φ_μ used in the LCAO expression above should better be already **adapted** or **deformed** to the molecular scenario because the molecular

---

2 Unfortunately, there are actually **two** types of atomic units, Hartree and Rydberg.
3 That is fine for atoms but exceptions exist for certain molecules (transition-metal complexes) whose HOMO–LUMO gap is small.

potential differs from the atomic one. And because of that, it is customary to drop the term "atomic orbital" in the LCAO context and replace it with the more general atom-centered **basis function**, representing the part (atom) within the system (molecule). Either way, it is a one-electron function belonging to or centered on a particular atom.

The antisymmetry of the resulting wave function, required by the fact that we are dealing with electrons that are fermions, half-spin particles, would be guaranteed by a so-called Slater determinant over all one-electron wave functions; the Pauli principle is in control. This corresponds to the Hartree–Fock (HF) approach (Fock, 1930a; Fock, 1930b) correctly including all **exchange** interactions[4] between the N electrons:

$$\Psi(r) = \frac{1}{\sqrt{N!}} \begin{vmatrix} \psi_1(r_1) & \psi_2(r_1) & \cdots & \psi_N(r_1) \\ \psi_1(r_2) & \psi_2(r_2) & \cdots & \psi_N(r_2) \\ \vdots & \vdots & \ddots & \vdots \\ \psi_1(r_N) & \psi_2(r_N) & \cdots & \psi_N(r_N) \end{vmatrix} \tag{2.5}$$

This famous and highly successful HF strategy, focusing on the all-electron wave function, is considered an excellent choice for molecules that are not extremely large, and finding it corresponds to solving a Fock-style eigenvalue equation

$$F\psi_i(r) = \varepsilon_i \psi_i(r) \tag{2.6}$$

in which the Fockian operator F consists of a one-electron Hamiltonian h, a direct Coulomb term J_j, and an exchange Coulomb term K_j of purely quantum-mechanical origin mirroring the wave function's antisymmetry in combination with a two-body operator (Szabo & Ostlund, 1989; Helgaker et al., 2000; Mayer, 2003):

$$F = h + \sum_{j=1}^{N} (J_j - K_j) \tag{2.7}$$

While the one-determinant approach in eq. (2.5) is very flexible, the usual (simplest) choice for solution yields orthogonal, **canonical**, delocalized orbitals, and these canonical orbitals may be transformed into more local ones, if that is needed to extract more "chemistry" from the wave function.

Because the many-electron HF wave function still misses the energy of electronic correlation E^{corr} (regardless of their spin taken care by the Slater determinant, they still repel each other), its size may be quantified by the energy deviation from the exact energy E^{exact} using

---

4 Quantum chemistry purists will insist that there is no exchange interaction at all but only Coulomb, namely (a) a classical averaged Coulomb energy, (b) a correction to the many-particle correlated Coulomb energy, and (c) a quantum correction term to the Coulomb energy; the latter is usually called **exchange** energy. I will stick to the usual or casual use of language because it is more widespread.

$$E^{corr} = E^{exact} - E^{HF} \tag{2.8}$$

as formulated by Löwdin, and various ways to improve the HF result are possible, either by involving several Slater determinants corresponding to a different electronic occupation of the molecular orbitals by the electrons, the so-called (restricted) configuration interaction (CI) which is easily made transparent chemically (Coffey & Jug, 1974), or methods of perturbation theory according to Møller–Plesset (MP), or coupled-cluster theory (= CI made size-consistent but not variational). There is a very rich literature from molecular quantum chemistry, and the interested reader is referred to likewise excellent textbooks covering just that (Szabo & Ostlund, 1989; Helgaker et al., 2000; Mayer, 2003), if needed.

For large molecules, the aforementioned approach becomes cumbersome in terms of sheer computation (scaling with fourth order for HF theory, with fifth order or higher for correlated orbitals), so the smart quantum chemists were forced to sort out which integrals of HF theory were important enough to compute and which could be simplified or simply discarded, the birth of highly successful semiempirical molecular-orbital theory (Bredow & Jug, 2017) such as complete neglect of differential overlap (CNDO). The situation is even worse if we reach out for crystalline matter although HF theory even including periodic MP perturbation corrections can be carried out (Usvyat et al., 2017) for insulators and semiconductors. For other solid-state materials, in particular, **metals**, HF theory is not a good starting point at all since correlation is likewise important to begin with, so correlation **cannot be dropped** like in many simple molecules since the electronic density in solids is **higher**, and that is due to more dense packing (using an unrigorous argument). And because of that, running HF theory with periodic boundary conditions (see below) to model a crystal may not help at all whenever metallicity is touched, then HF theory may spectacularly fail.[5] This is the main reason for the unfortunate schism between molecular and solid-state quantum chemistry. The majority of all elements showing up as solids is **metallic**, and one should not forget about that.

Hence, the solid-state theorists had to find yet another way to calculate the many-body electronic structure, and they did, thanks to the physicists (Inkson, 1986). It turns out that the total energy of whatever crystal (and also molecule and atom, thank you for that, too) with interacting electrons can be expressed as a **functional**[6] of the electron density

$$E\{\rho(r)\} = T_0\{\rho(r)\} + \int dr \, V_{ext}(r) \, \rho(r) + E_H + E_{XC}\{\rho(r)\} \tag{2.9}$$

which consists of just four ingredients after the equation sign: first, there is the kinetic energy of **noninteracting** electrons – hence, T_0 – of the same density as the interacting

---

5 The analytical HF ground state of a simple metal corresponds to a **zero** density of states at the Fermi level, a sheer theoretical disaster.

6 A functional is a function of a function. So, the functional of the electron density is called as such because it depends on the density which itself is a function of space.

electrons, second the Coulomb potential due to the nuclei, third the simple Hartree[7] term, and fourth an exchange-correlation term for correcting the grossly false assumption that the electrons have been formulated as being noninteracting. This is the incredibly tricky way density-functional theory (DFT) works, and its success story is easy to understand **a posteriori**. First, $T_0\{\rho(r)\}$ is pretty large and can be straightforwardly (i.e., second derivative) calculated given a set of one-electron wave functions (orbitals), and, second, the correction term $E_{XC}\{\rho(r)\}$ is often rather small, unless the material is too "correlated"; here, the physicists may think of transition-metal oxides, for example. Because of that, Schrödinger's many-body equation working with $\Psi(r)$ gets rewritten into sets of one-electron Kohn–Sham equations from the mid-1960s (Hohenberg & Kohn, 1964; Kohn & Sham, 1965) working with $\psi_i(r)$

$$\left[-\frac{\hbar^2}{2m}\nabla^2 + V_{\text{eff}}(r) \right] \psi_i(r) = E\,\psi_i(r) \tag{2.10}$$

in which an **effective** potential takes care of many-body effects given a reliable functional of exchange and correlation. Thanks to that, we are back at a one-electron theory allowing for easy interpretation, and the many-body effects have been kindly smuggled into that framework, a kind of pimped-up Hartree theory. In fact, that idea was somehow obvious because Slater (1951) had recognized quite early that the many-body nonlocal exchange potential of HF theory can be approximated from the cubic root of the electron density, the key to DFT.

Before considering this truly grand idea of electronic-structure theory (Parr & Yang, 1989) more closely in terms of the exchange-correlation term, it is probably a good move to provide further literature to the interested reader, in particular because DFT has totally reshaped both computational physics and chemistry, and we may safely consider it **the** workhorse of today's theorist. Both the DFT origins (Jones & Gunnarsson, 1989) as well as its tentative future prospects (Jones, 2015; Springborg & Dong, 2017) have been competently sketched. In fact, DFT has become so overwhelmingly present in daily work that it has allowed many practitioning scientists to chat about its pluses and minuses (Teale et al., 2022) including a wealth of up-to-date references.

Coming back to the decisively important exchange-correlation term, for $E_{XC}\{\rho(r)\} = 0$, one is still stuck in "Hartree hell." The **exact** functional is unknown, of course, but the local-density approximation (LDA) with

---

[7] The Hartree approximation is the basis behind eq. (2.3), and here we assume that each electron moves in a potential "sea" generated by all the other electrons. Because each electron belongs to the same "sea," accurately calculating the potential corresponds to iterating toward self-consistency until the incoming and outcoming solution no longer differ. That being said, one may prove that the self-interaction energy of a Hartree electron is zero in the limit of a large system (Inkson, 1986).

$$E_{XC}^{LDA}\{\rho(r)\} = \int dr\, \rho(r)\epsilon_{XC}\{\rho(r)\} \tag{2.11}$$

already gives surprisingly accurate results in many cases of slowly varying densities although that was unexpected at the very beginning. So, for each point in space, some amount of exchange and correlation (tabulated from an electron-gas calculation of the same density) is added. To improve further, an approximated generalized gradient-corrected functional (GGA) of the form

$$E_{XC}^{GGA}\{\rho(r)\} = \int dr\, \rho(r)\epsilon_{XC}\{\rho(r), \nabla\rho(r)\} \tag{2.12}$$

will prove useful, and then the gradient of the density also enters the functional expression. This may be followed by a meta-GGA functional (involving more than the first derivative), followed by a hybrid functional containing exact exchange (HF theory is entering, once again), and so forth. Right now, there is a **plethora** of density functionals available giving DFT an almost semiempirical touch but, in practice, a rather restricted number is actually being used. And it depends on the community. Molecular quantum chemists (Koch & Holthausen, 2001) may prefer the somewhat empirically looking but well-performing B3LYP functional while quite a few solid-state theoretical physicists still use the LDA: simple but transparent. When it comes to getting the relative stabilities of (magnetic) transition-metal allotropes right, the GGA is a must for physics, however (Dronskowski, 2005). It is virtually impossible to come up with a good recommendation (without being killed by parts of the community), so a proper reference (Teale et al., 2022) may suffice.

How about the solid state in particular? We alluded to that already, DFT is a child of the solid state (and metal physics), and because most solid-state materials present themselves as crystalline matter, one may benefit from essentially the same mathematical calculus as permanently used by the crystallographers dealing with crystals, so the introduction of **reciprocal space** is needed. Given the latter, one may come up with a new quantum number dubbed k, both fractional and directional, and one may translate any wave function fulfilling a Schrödinger or Kohn–Sham equation by a translation vector T within the crystalline structure. Alternatively expressed, the entire electronic structure of a translationally invariant solid can be brought back (or "backfolded") into the unit cell of reciprocal space, the Brillouin zone. This most important theorem of solid-state science (Bloch, 1928) was first formulated by Bloch, and it looks refreshingly simple but has enormous consequences:

$$\Psi_k(r+T) = \Psi_k(r)\, e^{ikT} \tag{2.13}$$

Instead of calculating a system composed of, say, 1 mol (about 6×10^{23}) of atoms, the Brillouin zone will suffice, thanks to translational symmetry, what an accomplishment. So, any solid-state electronic-structure calculation of crystalline matter therefore samples its results at various points of k space, the price paid for making an infinite system somehow "finite," and the exponential prefactor is decisive. It plays a similar role like the $c_{\mu i}$ LCAO mixing coefficients in the molecular case (eq. (2.4)). The idea is schematically sketched (Deringer & Dronskowski, 2013) in Figure 2.1.

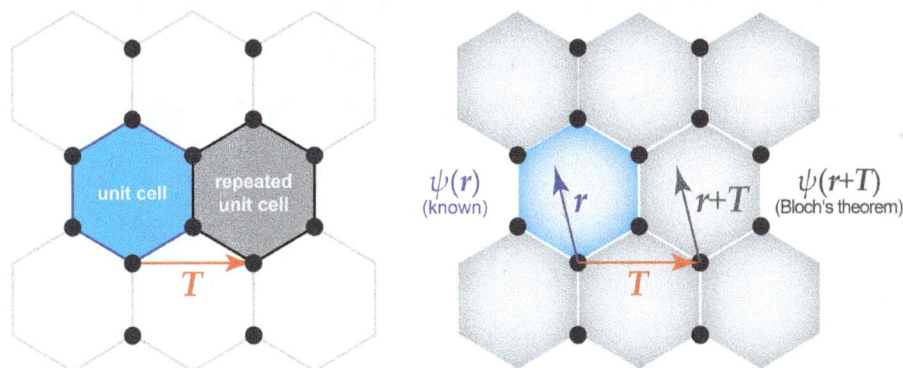

Figure 2.1: Sketch of a unit cell and its repeated unit cell in crystalline matter (left) and the consequence for the wave function and its translated function by using Bloch's theorem (right).

For completeness, we add that sampling reciprocal or $\boldsymbol{k}$ space nowadays is a routine business, and various ways to do so exist, for example, by selecting the "best" points or the most effective grid (Chadi & Cohen, 1973; Monkhorst & Pack, 1976), then followed by the most effective integration (Blöchl et al., 1994). More details are to be found elsewhere (Schwarz & Blaha, 2017); they look rather technical (albeit important) to me.

As alluded to already, translating a wave function is equivalent to multiplying it by the phase factor e^{ikT}, nothing more, which **really** simplifies things. Bloch himself was rather delighted to find by a simple Fourier analysis that the wave function differs from a plane wave of a free electron only by a periodic modulation. And the periodic modulation e^{ikT} mirroring the translational invariance of crystalline matter has important consequences for the basis functions to be used. This very periodic modulation means that plane waves serve as **symmetry-adapted** wave functions to a crystal's boundary conditions, so any periodic wave function may be expressed as a sum over exponential (or sine, or cosine) functions, totally delocalized to begin with, up to a certain lattice vector $\boldsymbol{G}$:

$$\Psi_k(\boldsymbol{r}) = \frac{e^{ikr}}{\sqrt{\Omega}} \sum_{G=0}^{\infty} c_k(\boldsymbol{G}) e^{iGr} \qquad (2.14)$$

This yields enormous mathematical advantages, for example, in terms of a simple control knob for the basis-set **quality** (measured by a maximum energy for the plane waves) and as regards exact interatomic Hellmann–Feynman forces (also known as nuclear gradients). The maximum cut-off energy is expressed via

$$\frac{\hbar^2}{2m}|\boldsymbol{k}+\boldsymbol{G}|^2 \le E_{\mathrm{cut}}. \tag{2.15}$$

On the other side, only valence electrons can be efficiently treated since the strong nodal behavior of the core-like orbitals below the valence orbitals would require an exceedingly large plane-wave basis set. This is most easily illustrated from a simple model system, a one-dimensional crystal made up of Na atoms.[8] At the zone edge, namely the special point **X** of the Brillouin zone, the extended wave function will look like Figure 2.2.

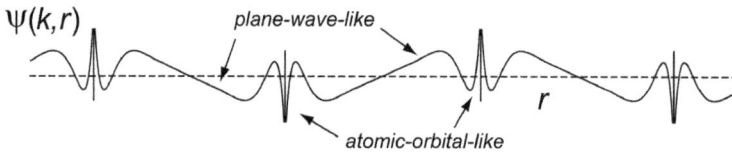

Figure 2.2: Extended wave function formed by 3s orbitals of a one-dimensional crystal made up of Na atoms at the **X** point of reciprocal space.

The wave function is sine- or cosine- or plane wave-like in character, moving up and down, positive and negative, as we go from unit cell to unit cell (or atom to atom), resulting from the translational invariance, and the zone edge = the **X** point of reciprocal space ($k = \pi/T$, which makes the wave function change sign). Close to the atoms, however, the wave function shows stronger oscillations; it becomes **nodal**, simply because the 3s orbital is orthogonal to the lower-lying core orbitals 2p, 2s, and 1s. And if we had to model that nodality, we would need quite a large number of plane waves up to a very high energy, so this would quickly kill plane-wave theory (Dronskowski, 2005).

Hence, **pseudopotential** (or effective core potential) theory is needed, replacing the true nuclear potential by a soft potential for the valence electrons and throwing away the entire atomic core. Why is that? Any valence electron is not only attracted by the nucleus but also pushed away by the lower-lying core electrons in-between nucleus and valence electron. Hence, only a weak pseudopotential is left (Hellmann, 1935; Hellmann, 1937a; Hellmann, 1937b). This strategy – which first came to the mind of Hellmann[9] and which corresponds to transferring the idea of the periodic table to quantum mechanics – is both justified and critical as there are many (in fact, infinite) ways to do that. The pseudopotential actually used may be norm-conserving (yielding the same density), energy-conserving (same eigenvalues), shape-conserving (same wave function after some distance from the nucleus), and so forth. This arbitrariness

---

8 The crystallographers insist – for good reasons – that there are only three-dimensional crystals but the theorists think that one-dimensional crystals serve nicely in theory.

9 Hellmann called it "combined approximation method," and the "pseudopotential" term used nowadays was coined later when Hellmann's findings were rediscovered after World War II (Jug et al., 2004).

is a problem, at least from the perspective of those who are in favor of all-electron calculations.

At the present time, the so-called projector augmented wave (PAW) theory is probably the best pseudopotential theory because the valence part somehow "rests" on a previous all-electron calculation (Blöchl, 1994). In fact, PAW theory is a happy marriage between pseudopotential theory and linearized augmented plane wave (LAPW) theory, arguably the most precise all-electron band-structure method ever invented (Andersen, 1975; Schwarz et al., 2002). For each atom situated inside a certain atomic sphere, PAW pseudopotential theory defines the true and nodal wave function $|\psi_i\rangle$ of the valence levels as

$$|\psi_i\rangle = |\tilde{\psi}_i\rangle + \sum_{\mu,R} c_{\mu R}\left(|\phi_{\mu R}\rangle - |\tilde{\phi}_{\mu R}\rangle\right) \tag{2.16}$$

and it is composed of a pseudo-wave function (first entry after the equation sign) inside the same sphere purely expressed by plane waves plus a correction term (second entry) which consists of all-electron and pseudo-"partial waves" (this is how physicists call numerical orbitals or basis functions). So, the pseudo-wave function is attached to the all-electron orbitals, as said before; hence, PAW pseudopotential theory gets rid of most of the arbitrariness mentioned before. Details are found elsewhere; they are somewhat tricky (Goedecker & Saha, 2017).

And so the story goes. For a given solid-state material, the Kohn–Sham equations are solved in reciprocal space using a basis of totally delocalized plane waves. To allow for that, only the valence orbitals are being used, given that their (nuclear) potential has been elegantly redefined by a pseudopotential (or PAW) theory. All the interactions between the electrons are taken care of by DFT in its various dialects (i.e., exchange-correlation functionals) fitting the problem. For a truly extended material, the unit cell is either the crystallographic unit cell, a smaller primitive unit cell derived from that, or a larger supercell. The latter may also be used if a molecular object is being studied, by allowing for enough empty space around it. And then the calculation is done over and over again, not only to achieve self-consistency but also at various k points of reciprocal space, to sample its properties. This procedure yields decent densities (needed for an improved potential) and densities of states. Alternatively, one may look at how the self-consistent energy levels change while walking along certain paths within the reciprocal space, and these **band-structure calculations** gave the method its name.[10] Based on a well-converged electronic structure, that is, converged in energy as a function of k space sampling and variational freedom (size of basis set as measured by a certain energetic cut-off for the plane waves), one may then proceed and calculate interatomic forces to derive (quasi-harmonic) phonons, the most impor-

---

10 Rumor has it that some people nowadays performing sophisticated DFT simulations on solid-state materials are not necessarily aware of the fact that they are actually doing band-structure calculations.

tant ingredient to get to free energies at finite temperatures, such as the Gibbs free energies, from which thermodynamic properties may be elegantly derived (Stoffel & Dronskowski, 2017), unless too many imaginary phonons are present, signs of instability. But that is quite another theoretical topic not to be explained any further.

There are virtually countless treatises on how to perform such electronic-structure simulations of perfect (and defective) solid-state materials, even from a profoundly chemical perspective (Bredow et al., 2009), and there are even entirely different approaches **not** starting from the one-electron but two-electron picture (Plekhanov & Tchougréeff, 2017). Likewise, there is no room here for covering excessively large simulations breaking the computational limit of DFT, so the chemists had to find a way to couple quantum mechanics (for the interesting, "reactive" part of the system) with molecular mechanics (for the less interesting hull around), and such methods exist as well (Catlow et al., 2017).

So far we have dealt with electronic-structure theory, at least in principle. The "only" problem left is to come up with a good chemical composition (which atoms are involved?) and a good structural model (where are those atoms?), then the calculation may start. The structural model is particularly important, often rather trivial when dealing with simple molecules such as hydrocarbons because we know the "sticks" connecting the atoms, even if the molecule has not been synthesized: C has four "sticks," N three, O two, and hydrogen one. Simply because of that it is quite easy to come up with a reasonable structural model in such molecular cases.[11] For an unknown solid-state material, say, an intermetallic compound containing 3 Pu, 11 Hf, and 7 Cr atoms (I have no idea if such a compound exists), the structure may be entirely unknown because nobody knows how many "sticks" connect Pu to Hf, Hf to Cr, and Pu to Cr. This problem – the unsolved valence problem of (solid-state) inorganic chemistry – makes this science more complicated than (molecular) organic chemistry **by orders of magnitude** but nobody wants to confess that in public.

Hence, one may try (= is forced to start with) a plethora of different structural models, and then select and cross-breed and multiply them, using evolutionary or particle-swarm or related algorithms (Oganov, 2010; Wang et al., 2010; Lonie & Zurek, 2011; Yu et al., 2017), such as to have a good representation of configurational space, followed by an energetic ranking of all the candidates. Hopefully, the material looked for is the one with the lowest energy, the ground-state structure, but other structures may be important, too. This point cannot be overstressed.

Although this chapter deals with plain theory, a word of caution is needed, and because it is so extraordinarily important, it really belongs here, not in the Appendix. Often **stable** new compounds found by sheer theory are routinely identified by their total energies being **lower** than those of the competing phases (starting materials or

---

11 The difficulties come later when such "simple" molecules change shape, generating problems as regards conformational sampling.

decay products); in simplified chemical language, these are **exothermic** compounds.[12] And if their phonon spectra are devoid of imaginary modes, the theorists are particularly happy because they phrase those materials as **mechanically** stable, an additional stability criterion. All those compounds that are not exothermic but **endothermic** (their energies are higher, not lower), however, are deemed unlikely or even impossible.

While that sounds theoretically (or physically) correct, it is plainly wrong chemically, that is, in the real world. Chemists are particularly good in making **endothermic** (unstable) compounds by tricking thermodynamics through kinetics. Almost the entire **organic** chemistry deals with unstable (endothermic) compounds that should spontaneously decompose into the thermodynamic sinks CO_2 and H_2O – but these molecules (and you and me, too) exist, thank God, due to large enough **activation barriers** against decomposition. Chemists call those compounds **metastable**. Likewise, the **inorganic** molecule hydrazine, N_2H_4, endothermic (unstable) by +159 kJ mol^{-1} can be made and is perfectly manageable (metastable), and even the solid-state explosive lead azide, $Pb(N_3)_2$, can be made quantitatively, given a lot of caution – until you touch it too hard, then it releases 556 kJ mol^{-1} with a loud bang and forms Pb and N_2. The aforementioned energy criterion would miss all those **interesting** compounds, far more interesting than good ol' boring NaCl and MgO, thermodynamic sinks.

In addition, many of the metastable compounds **do** show imaginary phonon modes. While this irritates the theorist, nature stays calm and lets such metastable compounds decay, either rapidly or slowly over many years. In the latter case, there is ample time for doing more chemistry with metastable compounds. Let us not forget that the presence of metastable phases and their tendency to crystallize more quickly than the stable phases is known since the times of Ostwald, hence Ostwald's step rule minimizing entropy production (van Santen, 1984). Sometimes it takes several decades until a compound formerly considered "stable" appears to be only metastable because an energetically even lower polymorph is accidentally found. The case of copper azide, β-CuN_3, exhibiting heterographene-like layers (Liu et al., 2015) is a good example.

That being said, future theorists are well advised not to overestimate the aforementioned energy (or thermodynamic) criteria but to **think about kinetics**, too, in particular about activation barriers. Existing compounds are not necessarily exothermic but they exist because of sufficiently large activation barriers against decay. There is a lot of work to be done theoretically!

---

12 In non-chemistry communities, such total-energy differences are often given in the units of eV atom^{-1} or Rydberg atom^{-1}. While this eases life for the computational scientist, it lacks any **chemical** meaning because it does not relate to **molecules** or **formula units**. Those who really want to collaborate with synthetic chemists in order to eventually make things should convert to kJ mol^{-1}, irrespective of whether molecules (say, OsO_4) or extended materials (say, MnO_2) are meant.

3 Analyzing molecules and solids

For hard-core theorists stemming from computational physics or materials science, it may be puzzling to realize that various ways of understanding or explaining or at least "phrasing" the atomistic world have been developed **without any knowledge of quantum mechanics**, successfully so. Even if the reader is an expert in the quantum-mechanical calculation of molecules and solids, glimpsing into what follows may be truly worthwhile, at least I hope so.

To begin with, the periodic table of the elements (PSE), ingeniously put together in the decade of the 1860s by Dmitri Ivanovich Mendeleev (Mendelejeff, 1869) – and let us not forget the contributions of Lothar Meyer (1864) – is a fascinating example. By observing certain trends in the behavior of chemical elements, one may put those elements into certain "groups" of chemical **similarity**, which simply reflect their valence-electron counts and configurations (using modern language). And by doing so, one may directly understand or explain chemical reactivity as a function of the electron count and orbital occupancy (using modern language, once again), so alchemy really turned into chemistry. And because of that, at the end of the nineteenth century one could **rationally** find new elements (Ge and Ga, for example) and hand them over to the physicists to allow them to perform careful measurements, thereby looking for atomic properties, so-called observables, from which quantum mechanics could be derived. At the same time, the PSE story tells us that one needs a good theory to derive observables from experiment, so there is some kind of theory-loadedness of observations, at least in **real** science, which cannot be avoided. I trust that somebody will eventually find a good excuse for the embarrassing fact that Mendeleev was never awarded the Nobel Prize in Chemistry.[1]

A second example for chemical imagination is the covalent bond and its graphical formulation, another groundbreaking step in the history of chemistry, and here the credit goes to, among others, Gilbert Newton Lewis (1916). Lewis basically formulated the "cubical atom" such as to distribute a nonmoving electron **octet** (see Appendix A) around a main-group atom, and he then realized that a pair of electrons connecting two atoms could be graphically represented by a "dash" between them, the "single" atom–atom bond. Up to the present day, we are still using the ingenious "Lewis structure" (*Valenzstrichformel* in German) when writing down the molecular formulas of the hydrogen molecule, H–H, the ethylene molecule, $H_2C=CH_2$, and any protein of

---

[1] Nonetheless, he and Meyer received the Davy Medal in 1882.

https://doi.org/10.1515/9783111167213-003

whatever size. And the method also works for heavier atoms, transition metals, and so forth, given a few improvements. Science historians will need to find out why Lewis, too, never got a Nobel Prize. As our chemical thinking has evolved from the chemical behavior of an atom and its valence-electron count (Mendeleev) and the bonding proclivity as expressed through a "valence formula" also alluding to the molecule's geometrical shape (Lewis), we are very well advised to search for their correspondences in electronic-structure theory, and it would be simply foolish not to do that. So, we need to find out what can be done.

Hence, let us look at the famous chemical-bonding triangle by van Arkel (1956) and Ketelaar (1958) that has been around since the late 1950s, depicted in Figure 3.1. This very triangle is an incredibly powerful icon of chemical understanding, and it only relies on the electronegativity (EN) which, according to Pauling, is the ability of an atom in a molecule to pull electrons toward it (see Appendix B). For the x-axis, an average EN characterizes the system (a molecule or a crystal), whereas the y-axis shows the EN difference in that system (Deringer & Dronskowski, 2013).

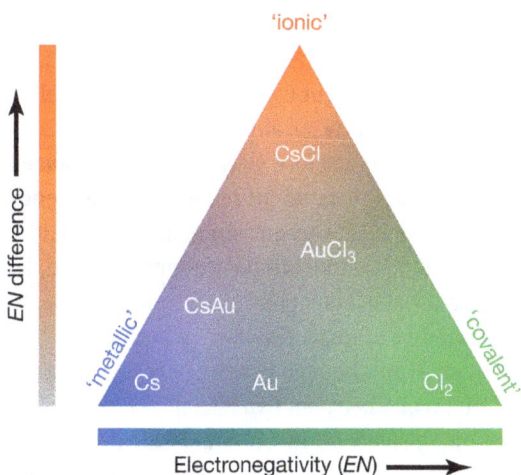

Figure 3.1: The famous van Arkel–Ketelaar triangle of chemical bonding that groups all chemical species by their EN differences (y-axis) and the average EN of atoms involved (x-axis).

Starting with a simple alkali halide such as CsCl (or NaCl, KBr, etc.) brings us directly to the area of **ionic bonding** or **ionicity**. For such materials, the EN difference is so large that an electron jumps from the atom with low EN (lab jargon: "electropositive" but that is really jargon) to the atom with a high EN, hence – in the case of CsCl – we are left with a Cs^+ cation and a Cl^- anion. Because of that, the formerly $6s^1$ valence electron of cesium finds itself caught in a $3p$ orbital belonging to chlorine, so Cl^- now has a noble-gas shell configuration (a full octet, $3s^2\,3p^6$, just like in the [Ar] case) and the same holds true for Cs^+ (likewise full octet, $5s^2\,5p^6$, [Xe] configuration). Note that the very electron that has been transferred sits in an **atomic** orbital, the chlorine $3p$ orbital, a kind of **one-center** situation. That being said, both cation and anion no longer "communicate" quantum-mechanically[2] but may be approximated by simple point charges, and **this** is the reason why everything else may be calculated using classical electrostatics, for example, the existence of a significant Madelung or lattice energy associated with such electrostatically bonded solids. The higher the charges and the smaller the distances between cation and anion, the larger the Madelung or lattice energy (slightly smaller, please see Appendix C). If you ever wondered why ionic compounds are covered in elementary solid-state chemistry books, **this** is the reason. Quantum mechanics is apparently not needed to begin with, unless one is interested in the correct cation–anion distance, then a Born correction term going back to Pauli repulsion between the filled atomic orbitals is required. In any case, understanding is not hampered by this huge simplification of turning quantum mechanics into classical electrostatics. Quantum mechanics is **certainly** required, however, if we want to know the non-observable atomic charge on cation and anion. Approximating cesium as a singly charged cation isn't too bad, but an oxide anion, formally O^{2-}, clearly deviates from two transferred electrons. We will come back to that somewhat later, also covered in Appendix C.

An entirely different world is given in the lower right corner of Figure 3.1, dealing with highly electronegative atoms such as Cl (or H, or N, or any other nonmetal atom). If at least two such atoms approach each other, one or more electrons may be **shared** between the two atoms, and this leads to the realm of **covalent bonding** or **covalency** ingeniously tackled only one year after the invention of Schrödinger's equation by Heitler and London.[3] Because covalency is a profoundly quantum-mechanical phenomenon, let us take a closer look needed to get rid of the usual oversimplifications sometimes found in other texts. First, **no electron pair is needed** to invoke covalency because the hydrogen molecular cation, H_2^+, is already covalently bonded despite only one electron shared. Second, **it is not the sharing of electrons** between two nuclei – an essentially classical idea – which leads to covalency but, instead, **quantum-mechanical interference between two wave functions** (Frenking, 2022), typically an orbital on the left and an orbital on the right atom. This interference leads to **electronic exchange**, something unknown in classical mechanics, and the new wave function lowers in energy; chemical bonding, including covalency, is an energetic phenomenon. For the simple hydrogen molecule H_2, there are two ways to model the resulting wave function, either by emphasizing the repulsion between the electrons, so-called electronic correlation, the original Heitler–London (1927) or valence-bond model, or the non-correlated model often associ-

---

2 Dear theorists, yes, I know. Both atoms still communicate but the electron jump has happened already, and this is the decisive step, at least in my eyes.
3 Schrödinger was not interested in chemistry, so his two postdocs did the job properly (Gavroglu & Simões, 2012). And they also understood that sharing electrons (= classical physics) is **not** decisive but the **interference of wave functions** is.

ated with Hund and Mulliken (Condon, 1927), also called molecular-orbital model. The latter, a more delocalized approach, is more common these days (and it has turned out more successful, if I may say so), but before we consider H_2 in more detail, let us look at the lower left corner.

The lower left corner in Figure 3.1 is found for Cs and other phases with a very small number of electrons and low electronegativities to begin with, the metallic elements and metallic compounds showing **metallic bonding** or **metallicity**. For most of them, their number of available electrons is actually so small that covalent bonding (= interference of wave functions) can certainly not happen through electron pairs (or single electrons); for example, Cs would need to have eight electrons to build eight two-center two-electron single bonds to its eight direct neighbors in the *bcc* crystal structure but it has only one electron. Hence, some kind of electronic socialism sets in, and **any** valence electron is totally delocalized over all the atoms, in particular for simple metals in which the electron is "free," a term from solid-state physics, the so-called Sommerfeld model (Ashcroft & Mermin, 1976). For a metal involving $3d$ or $4f$ electrons, the situation gets a little more complicated but, nonetheless, the delocalization of the electrons over all atoms – this also guarantees metallic conductivity – is almost the same. And since many (a plethora, actually) atoms are involved, we have some kind of multicenter covalency, exactly that, induced by an insufficient number of electrons. Any bonded electron is neither situated in a one-center orbital (ionic case), nor in a two-center orbital (the "normal" covalent case), not even in an orbital covering three, four, five, etc. atoms (like in some molecules) but in a **multicenter** orbital or **delocalized electronic band** (metallic case), simply speaking. That being said, metallicity is a special case of covalency taking into account that the number of electrons is far too low and the number of atoms far too large to allow for "normal" covalency. In addition to that, electronic correlation is very important in metals, too, more important than exchange, as witnessed from the failure of Hartree–Fock theory in metals (see Chapter 2).

Since we are interested in the quantum-chemical consequences and their analyses, let us now look a little closer at the hydrogen molecule, the quantum chemist's drosophila for explaining **orbital interactions**, chemists really like to think this way (Albright et al., 2013). In the case of H_2 and its associated H–H single bond, a molecular orbital (MO) would read

$$\psi = c_1\phi_1 + c_2\phi_2 \tag{3.1}$$

in which the left (1) and right (2) atoms would each contribute a $1s$ atomic orbital (ϕ) such as to generate an MO (ψ). This is the simplest case of the method dubbed **linear combination** of (adapted)[4] **atomic orbitals** toward **molecular orbitals** (LCAO-MO) already

---

4 To arrive at a variationally better MO of lower energy, the atomic orbitals should be **adapted** to the molecular potential. In the present case of H_2, the nuclear charge is twice as large as in atomic H, to be taken into account by a more **contracted** atomic orbital for each H atom (see below).

met in the previous chapter, and the resulting low-energy bonding orbital dubbed ψ_+ (or σ_g, with a σ bonding character and **gerade** = even inversion symmetry) has the same mixing coefficient, $c_1 = c_2$, on the left and right atoms, namely $c_1 = 1/\sqrt{2(1+S_{12})}$ in which $S_{12} = \langle \phi_1|\phi_2 \rangle$ is their overlap integral. The other MO, much higher in energy due to its antibonding character and called σ_u^* (still σ but **ungerade** = odd inversion symmetry, the asterisk indicates antibonding) is defined by $c_1 = -c_2$. The MO diagram looks like the one given in Figure 3.2.

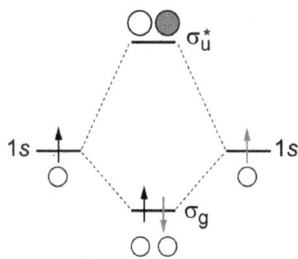

Figure 3.2: The molecular-orbital picture of H–H bond formation in the H_2 molecule, starting from two $1s$ atomic orbitals and arriving at a bonding σ_g and an antibonding σ_u^* molecular orbital.

The important point to understand is that the bonding and antibonding characters of the two wave functions are simply determined by their plus–minus signs of the LCAO coefficients or, in other words, by their **phases**. Both a plus–plus and a minus–minus combination, $c_1 = c_2$, will yield a bonding MO whereas a plus–minus combination, $c_1 = -c_2$, stands for an antibonding interaction.[5] And this is of **paramount importance** because the wave function **itself** carries that bonding–antibonding information, to be found by observing how two orbitals (or more) yielding that wave function interfere with each other (Dronskowski, 2005). And then it is also trivial to understand that the electron density (which is always zero or larger than zero but never negative) does **not** contain, under no circumstances, that precious information – unless the phases are recovered, so to speak, by an orbital basis leading to exactly that density. **It all depends on the phases,**[6] as was made crystal clear already (Frenking, 2022).

To make the H_2 story complete, the low-lying σ_g MO reads $\psi_+ = \frac{\phi_1 + \phi_2}{\sqrt{2(1+S_{12})}}$, and its electronic energy is $E_+ = \frac{H_{11} + H_{12}}{1 + S_{12}}$ in which $H_{11} = \langle \phi_1|H|\phi_1 \rangle$ and $H_{12} = \langle \phi_1|H|\phi_2 \rangle$ designate expectation values of the Hamiltonian operating on both molecule-adapted atomic orbitals. Given an H–H distance of 0.74 Å, the S_{12} overlap integral arrives at ca. 0.64 when calculated with a $1s$ Slater orbital and a $\zeta = 1.3$ contraction parameter, not 1.0 like in the H atom itself (because the nuclear charge is larger, see above). And now we may square-

---

5 This is only true because two s atomic orbitals are involved. For a combination between an s and a p atomic orbital, the overlap integral (and the type of interaction) not only depends on the distance but also on the **symmetry** (mutual orientation) because the two lobes of the p atomic orbital differ in phase.
6 And that immediately explains why some people express doubts when it comes to analyzing chemical bonding **only** with the density, like in Bader's AIM approach (1990), because the phases are entirely missing.

integrate the bonding MO over the entire molecule (please recall that all involved orbitals are real, and complex conjugates are irrelevant) and arrive at

$$\int \psi^2 d\tau = c_1^2 \int \phi_1^2 d\tau + c_2^2 \int \phi_2^2 d\tau + 2c_1c_2 \int \phi_1\phi_2 d\tau \tag{3.2}$$

or in bra-ket notation

$$\langle \psi | \psi \rangle = c_1^2 \langle \phi_1 | \phi_1 \rangle + c_2^2 \langle \phi_2 | \phi_2 \rangle + 2c_1c_2 \langle \phi_1 | \phi_2 \rangle \tag{3.3}$$

which is a rather complicated way to write $1 = 0.305 + 0.305 + 0.390$, easily shown (do it yourself!) if we insert $\langle \phi_1 | \phi_2 \rangle = S_{12} = 0.64$ into eq. (3.3) and reconsider that atomic and molecular orbitals are normalized to unity. And if we insert two electrons into the low-lying σ_g MO, the same differential calculus is multiplied by 2, and then it simply reads $2 = 0.61 + 0.61 + 0.78$. We may thus conclude that 0.61 electrons are on the left H atom (its **net population**), the same is true for the right atom by symmetry, and there are 0.78 electrons (the **overlap population**) between the atoms. Welcome to Mulliken population analysis (1955a; 1955b), a decisive step toward chemical understanding. The so-called **gross population** (GP) of either left or right H atom is achieved when we distribute the overlap population evenly ($0.78/2 = 0.39$) between left and right atoms (strictly valid only if they are of the same kind) and add it to the net populations, so this reads $0.61 + 0.39 = 1.00$ for both H atoms, as expected. The atomic **charge** of either H as defined by Mulliken is given by its neutral electron count (identical to the nuclear charge) minus the GP, hence $1.00 - 1.00 = 0$. So, there is no charge transfer in H_2; both atoms are electrically neutral, as also expected.

For a positively charged H_2^+ molecule, the story remains essentially the same but one electron (either the gray or black one in Figure 3.2) leaves the scene, thus corresponding to the reaction product between one H atom and one proton. Because of the interfering atomic orbitals, there is still bonding, although a little weaker, and nature does **not** need an electron pair to secure a covalent interaction, as said before. Pairing is essential for biology, not for chemistry.

And all that may be generalized, of course. For a molecule of any size, no matter how big, the LCAO-MO construction simply contains (many) more terms ($2s$, $3p_x$, $4d_{xy}$, $5f_{yz^2}$, etc.) but the principle stays exactly the same such that any MO will read

$$\psi_i = \sum_{\mu=1}^{n} c_{\mu i} \phi_\mu \tag{3.4}$$

One may also group atomic orbitals to the atom they belong to but that only changes the notation, not the calculus. To cover the entire Hilbert space (and achieve maximum accuracy), the atomic basis set should be complete, so one is well advised to use entire sets of Slater-type orbital (STO) or Gaussian-type orbital (GTO) (Szabo & Ostlund, 1989), the usual suspects in molecular quantum chemistry.

When Mulliken population analysis is then applied for a complicated molecule, it reads

$$\int \psi_i^* \psi_i \, d\tau = 1 = \sum_\mu c_{\mu i}^2 \int \phi_\mu^2 d\tau + \sum_{\mu < \nu} 2 c_{\mu i} c_{\nu i} \int \phi_\mu \phi_\nu \, d\tau$$

$$= \sum_\mu c_{\mu i}^2 + \sum_{\mu < \nu} 2 c_{\mu j} c_{\nu j} S_{\mu \nu} \tag{3.5}$$

$$= \sum_\mu NP_\mu + \sum_{\mu < \nu} OP_{\mu \nu}$$

and now we have formulated **net populations** on orbitals, NP_μ, and **overlap populations** between orbitals, $OP_{\mu\nu}$. One can also add all orbital net populations on the same atom to an **atomic net population**, and one can define an **atomic gross population** abbreviated as GP_μ, so then the atomic charge q of an atom A with originally N electrons would read

$$q_A = N - \sum_{\mu \in A} GP_\mu \tag{3.6}$$

if we include all the orbitals belonging to atom A; nothing changes. Alternatively, one may invoke Löwdin's population analysis for which the basis set is orthogonalized by applying the corresponding symmetric orthogonalization, and the new orthogonalized basis functions in Hilbert space are "least distant" from the original non-orthogonal functions. The Löwdin approach (Szabo & Ostlund, 1989) shows some advantages here and there, for example, the orbital occupation is never smaller than 0 or larger than 2, unlike in the Mulliken scheme. And there are more variants, going by the names of their inventors, such as Davidson, Roby, Jug, and Ahlrichs, all trying to answer profoundly chemical questions.

Such population techniques of whatever specific method are incredibly powerful in the hands of a skilled quantum chemist, and they allow to **directly** analyze the wave function in terms of atoms and bonds, given that an orbital-based representation of the molecular wave function (such as the one given by LCAO-MO) is available. One must know the orbital **phases**, then the analysis may begin. And this is the reason why there are powerful computer programs specifically designed for **molecules**, all running with advanced basis sets composed of atom-centered functions, because this is the natural quantum-chemical **operational basis** to calculate and understand molecules in which atoms and their bonds are decisive.

Now, for solid-state calculations it is also possible to define the extended wave functions derived by using Bloch's theorem from local (adapted) atomic orbitals, and this approach is called the "tight-binding" approach in theoretical solid-state physics and also chemistry. For example, there is the good ol' tight-binding linearized muffin-tin orbital theory in the atomic-spheres approximation, TB-LMTO-ASA (Andersen & Jepsen, 1984), with Hankel functions, which was historically extremely useful to develop the COHP method (see below); a more general and also recent treatise on tight-binding theory is available (Seifert, 2017). As regards wording, "tight-binding" means that the electrons are tightly bound to the atoms, so the atomic wave functions (orbitals) are considered "good enough" to also serve in the periodic calculation.[7] If such methods are used, one may – just like for molecules – understand the material in terms of atoms and bonds. At present, however, we already learned that the vast majority of electronic-structure calculations for periodic solids are carried out by using **plane waves**, so the k-dependent wave function reads

$$\psi_{jk}(\boldsymbol{r}) = \frac{1}{\sqrt{\Omega}} \sum_{\boldsymbol{G}} c_{j(\boldsymbol{k}+\boldsymbol{G})}^{\mathrm{PW}} e^{i(\boldsymbol{k}+\boldsymbol{G})\cdot\boldsymbol{r}} \tag{3.7}$$

and while that has enormous computational advantages, not only in terms of efficiency but also as regards the principal absence of basis set bias or basis set superposition errors (hence accurate Hellmann–Feynman forces, **very** important these days to optimize structures), it is a true disaster for chemical interpretation since there are no atomic orbitals, so the entire "chemistry" (= interacting atoms) has vanished into thin air. We do also know, however, that exactly the same extended wave function can also be expressed by atomic orbitals as in

$$\psi_{jk}(\boldsymbol{r}) = \sum_{\mu\boldsymbol{T}} c_{\mu jk}^{\mathrm{LCAO}} e^{i\boldsymbol{k}\cdot\boldsymbol{T}} \phi_{\mu}(\boldsymbol{r}-\boldsymbol{T}) \tag{3.8}$$

the aforementioned "tight-binding" approach which is close to what is being done for molecules. Shouldn't it be possible to communicate between the two? That is of course possible (Edminston & Ruedenberg, 1963) and simply requires that we need to go back and forth between two different "pictures" of the same thing; in this case, the periodic wave function $\psi_{jk}(\boldsymbol{r})$. The same wave function can be correctly expressed by two very different bases, one that is totally delocalized (plane waves), the other one being totally localized (atomic orbitals), so we are talking about two different **representations** of the same quantum-mechanical wave function.

Changing between the two representations is equivalent to a unitary transformation between the two, and the "unitary" term (borrowed from geometry) means that the shape of the object (distances, angles) stays intact. Alternatively expressed, one deals with two equivalent coordinate systems of the same geometrical object.[8] In our case, the wave function should not change, only its **expansion** in terms of two different basis sets, delocalized and localized, must differ. Moving from the delocalized to

---

7 Hence, any standard atomic-orbital basis can be used, at least in principle. The CRYSTAL code, for example, allows to carry out tight-binding electronic-structure calculations for periodic solids, based on Gaussian orbitals, and it also allows for chemical-bonding analysis (Erba et al., 2022).

8 Whether an ant crawls on the surface of a shoe box or a soccer ball, its position is correctly described either by a Cartesian (x, y, z) or by a spherical (r, ϑ, φ) coordinate system. On the shoe box, the Cartesian system will be easier to deal with while on the soccer ball, the spherical system will excel.

the localized expansion of the wave function can be accomplished by **projecting** the delocalized wave functions to local orbitals (Chadi, 1977; Sánchez-Portal et al., 1995; Sánchez-Portal et al., 1996), mathematically spoken, and its chemical motivation could not be more straightforward (Deringer et al., 2011). The delocalized wave function rests on a good pseudopotential or PAW potential, and it is then projected to (or expanded by) a set of local orbitals that span the same Hilbert space because the object (the wave function) should not be distorted; otherwise, this would not be a unitary transformation. To do so, a so-called transfer matrix needs to be specified which, due to the rather difficult PAW theory, is far from being trivial, and its analytical elements cover both the "pseudopotential" and the "augmentation" part; details can be found elsewhere (Maintz et al., 2013; Maintz et al., 2016). The accuracy of the projection (or transformation) is easily specified by a so-called spilling parameter that measures how many electrons get "spilled" over occupied (charge spilling) or all levels (total spilling). The charge spilling is truly important, of course, and given proper reorthonormalization, no electron density gets lost despite a spilling parameter of typically a few percent.[9]

Now that the periodic plane-wave function has been exactly rewritten into a basis composed of local atomic orbitals, all the chemical-bonding tools originally derived for molecules as done by Mulliken and others are eventually available for analyzing whatever solid-state material, too. For example, the so-called crystal orbital overlap population (COOP), the solid-state equivalent of Mulliken's population analysis for periodic solids, is defined as a sum of energy-dependent orbital-mixing coefficients (= the density-of-state (DOS) matrix) and the overlap integral:

$$\text{COOP}(E) = S_{\mu\nu} \sum_{j,k} w_k \text{Re}(c^*_{\mu,jk} c_{\nu,jk}) \delta(\varepsilon_j(\boldsymbol{k}) - E) \tag{3.9}$$

Re indicates the real part. Hence, the DOS is weighted, so to speak, with Mulliken's overlap population, and a COOP plot will display both positive values (due to positive overlap populations from constructive interference = bonding) and negative values (negative overlap population, destructive interference = antibonding). If the overlap population is zero, the COOP is zero as well, indicative of a nonbonding interaction, so the two atoms or orbitals do not interfere at all. The COOP method, as simple as it may look, must be considered revolutionary because it eventually convinced the solid-state chemists (at least the smartest among them) to turn to quantum chemistry and abandon the simplistic ionic picture (see Appendix C) whenever possible (Hughbanks & Hoffmann, 1983).

---

9 This is in sharp contrast to the way standard plane-wave PAW-DFT often proceeds, say, for projecting local densities of states (DOS). For GaAs with $3 + 5 = 8$ electrons, the total DOS does yield 8 electrons up to the Fermi level but the sum of the local DOS will only yield 4.8 electrons since then just the electrons **within** the PAW spheres are added up, hence one misses in-between electron density. The LOBSTER program, however, will arrive at 7.998 electrons and even better than that because the projection is both analytical and exact.

COOP stems from semiempirical extended Hückel theory (Hoffmann, 1963), an enormously influential but non-variational quantum-chemical approach of the past, used for molecules and solids as well; it is yet another method belonging to the "tight-binding" zoo, probably the most popular one. Despite the many advantages of extended Hückel theory, in particular as regards understanding, it cannot compete with today's state-of-the-art methods in terms of accuracy.[10] With the advent of variational density-functional theory (DFT), an alternative tool was developed, dubbed crystal orbital Hamilton population (COHP), very analogous to COOP:

$$\text{COHP}(E) = H_{\mu v} \sum_{j,k} w_k \text{Re}\left(c^*_{\mu, jk} c_{v, jk}\right) \delta\left(\varepsilon_j(\boldsymbol{k}) - E\right) \tag{3.10}$$

Within the COHP method (Dronskowski & Blöchl, 1993), the corresponding elements of the Hamiltonian itself weight the DOS, and there are negative COHP (energy lowering from constructive interference = bonding) and positive COHP (energy increase from destructive interference = antibonding), so COHP and COOP often look like mirror images. To make COOP (semiempirical theory) and COHP (DFT) **look the same**, one simply plots **negative** COHP (so, −COHP) such that all bonding levels go to the right, now and in the future. Before it gets too complicated, we better show Figure 3.3 offering band structure, DOS, COOP, and also COHP analysis for elemental diamond, a simple example, and then everything is clear.

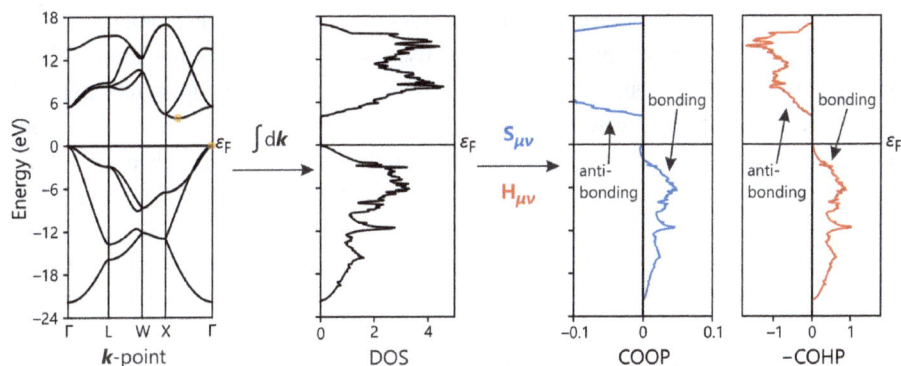

Figure 3.3: Electronic band structure of diamond (left), density of states (DOS, middle), as well as COOP and COHP chemical-bonding analysis (right). The highest/lowest points of the valence/conduction bands have been indicated.

The figure based on plane-wave DFT and a reasonable (GGA) functional is extraordinarily instructive; it holds **massive** amounts of information (Nelson et al., 2023). For

---

10 Its inventor once jokingly said that extended Hückel theory is a method to which all other methods are superior.

elemental diamond, the bands go up and down in energy as we walk through reciprocal space, and similar figures are well known from standard solid-state physics (or
theory) textbooks. One can even see that elemental diamond is an indirect semiconductor because the upmost regime of the valence bands and the lowest part of the
conduction bands do not occur at the same k point. By integrating over the entire Brillouin zone, an averaged measure, the so-called DOS, is gained which serves us as a
simple level counter;[11] if integrated over the energy, the correct number of electrons
(four valence electrons in this case) defines the position of the Fermi energy.[12] It is
obvious that the DOS contains **zero** chemical-bonding information because the orbital
phases are **not** involved, but in order to achieve that information, a COOP plot (operating with the overlap population between orbitals) or a COHP plot (operating with
the Hamilton population) delivers just that. And then one realizes that the entire valence band is bonding (only the slightly less precise COOP shows a small deviation
close to the Fermi level) and the entire conduction band is antibonding. As expected,
nature fills bonding levels and avoids antibonding ones. Note that we plotted the **negative** COHP to always have the bonding levels to the right and the antibonding levels
to the left, a convenient convention to avoid confusion, as written before. If one per-
forms a **projected** COOP or COHP calculation from plane waves, one sometimes
writes pCOOP or pCOHP but dropping the "p" is also fine; it is not needed, just a technical detail. In fact, the above figure was made by projection to a fine set of atomic
basis functions, by the way, and we simply dropped the "p" not to confuse the reader.

There is another, far more difficult and less precise way to elicit some bonding
information from the DOS, indirectly so. Because the phases are missing, one needs to
compare the energetic positions of the affected DOS levels **before** and **after** chemical
bonding has taken place; those going down/up are the bonding/antibonding levels. Because this requires a carefully prepared model and some detective skills to recognize
the relevant levels, it is very rarely carried out. In fact, COOP and COHP are more ele-
gant and straightforward.

It goes without saying that rewriting the entire delocalized electronic structure
into an orbital basis leads to unexpected, yet extremely helpful side effects in terms of
computing atomic charges. Now that all the orbital mixing coefficients are available
as a function of k space, an **orbital GP** is obtained:

$$GP_\mu = \sum_k \sum_v P_{\mu v}(k) S_{\mu v}(k) w(k) \tag{3.11}$$

---

11 By sheer convention, chemists often plot the energy along the y-axis and the DOS along the x-axis.
While this does not make physical sense at all (because the DOS depends on the energy, not otherwise), it helps to see how the DOS evolves from the band structure.
12 In fact, the Fermi energy is exactly in the middle between the valence and conduction bands but
chemists usually formulate it as the highest occupied level, that is, similar to the highest occupied molecular orbital (HOMO) in the molecular case. It is incorrect, strictly speaking, but convenient.

and, if summed up over all orbitals on that atom, also an **atomic GP** is available, hence, we know the **atomic charge** (according to Mulliken or Löwdin) without **any** additional effort; the wave function contains that precious information already because the phases enter the density matrix $P_{\mu\nu}$. And that is to say that Mulliken and Löwdin charges are available in plane-wave theory (Erturul et al., 2019), eventually, even l-resolved (i.e., for s, p, d, and f orbitals). This is in sharp contrast to a method dubbed atoms in molecules (AIM) in which the Bader atomic charge is derived from the density:

$$q_{A,\text{Bader}} = Z_A - \int_{V \in A} \rho(\mathbf{r}) \mathrm{d}\mathbf{r} \qquad (3.12)$$

This approach is popular in crystallography (they only have the density, not the wave function) and also in plane-wave theory because, until recently, a reliable projection to atomic orbitals was unavailable. Nonetheless, Bader's method (1990) takes at least an order of magnitude more time to calculate and sometimes even yields weird numbers, as later seen in a chapter dealing with battery materials. Because the unitary transformation is available, however, there is a powerful alternative to calculate atomic charges, in particular when an entire series of isostructural compounds with differing cations must be analyzed (Li et al., 2018a).

Before we focus on yet another solid-state bonding indicator, experience from molecular quantum chemistry comes in handy. Overlap and Hamilton populations look fine, but how about directly extracting the **bond order** from the molecular wave function? The bond order (single bond, double bond, etc.) is an essential part of chemical thinking since the times of Lewis, and it is indeed possible to derive it, quite easily so, because of the decisive work of Wiberg (1968) using an orthogonal basis set and Mayer (1983; 2017) using a non-orthogonal basis set, both focusing on the aforementioned density matrix $P_{\mu\nu}$, so the bond indices (BIs) read

$$\text{BI}_{\text{Wiberg}} = \text{Tr}\left|P_{\mu\nu}\right|^2 \text{ and } \text{BI}_{\text{Mayer}} = \text{Tr}\left(P_{\mu\nu}P_{\nu\mu}\right) \qquad (3.13)$$

and Figure 3.4 gives a simple example as derived from a straw-man organic molecule carrying the funny name (*E*)-3-(2-cyanovinyl)benzoic acid.

Clearly, C–C and C–O single bonds arrive at BIs of about 1.1, and the BIs (or bond orders) of the C–H single bonds are about 0.9, so the Lewis sketch is nicely recovered by the quantum-chemical tool. The same is true for the one C=C double bond (BI ≈ 1.7) and the aromatic C–C bonds in the hexagon[13] with BI ≈ 1.4, and even the C≡N triple bond arrives at BI ≈ 2.8. The small deviations indicate, however, that the simple human concept of single, aromatic, double and triple bonds, so-called **entities** (Appendix D),

---

[13] The hexagon-like core of the molecule is derived from the benzene molecule, C_6H_6, a so-called aromatic species, as explained in any organic chemistry textbook. The naming is due to the aromatic smell (yes, really) of benzene and similar molecules.

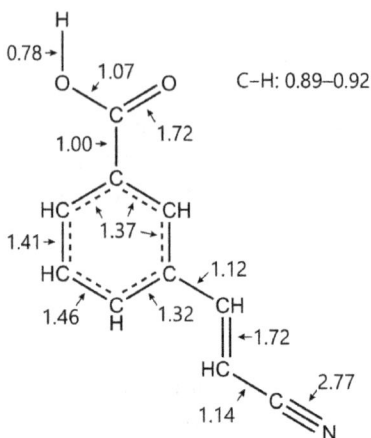

Figure 3.4: Bond indices for different chemical bonds in the molecule (*E*)-3-(2-cyanovinyl)benzoic acid, as calculated using LOBSTER.

does not fully cover the quantum-chemical reality, albeit it does so to a pretty large extent.

For the crystalline solid, an indicator targeted at the bond order and similar to COOP and COHP can also be derived, as alluded to already. Even the mathematical structure is the same but we now must replace the overlap integral $S_{\mu\nu}$ or Hamilton integral $H_{\mu\nu}$ by the corresponding element of the density matrix $P_{\mu\nu}$. The crystal orbital bond index (COBI) reads:

$$\text{COBI}_{\mu\nu}(E) = P_{\mu\nu} \sum_{j,k} w_k \text{Re}\left(c^*_{\mu,jk} c_{\nu,jk}\right) \delta\left(\varepsilon_j(\boldsymbol{k}) - E\right) \tag{3.14}$$

Another simple example will do, carried out using periodic boundary conditions and plane waves but for a single molecule (because it is so nicely simple) put into a large supercell. After having reached self-consistency, the chemical-bonding information can be projected. Figure 3.5 shows the electronic structure of the nitrogen dimer, N_2, and its N≡N triple bond visualized from (projected) DOS, COOP, COHP, and COBI.

To ease interpretation, the bonds and electron pairs in the Lewis structure of N_2 have been colored, so the σ single bond is in orange, the two π bonds are in red, and the two lone pairs to the left and right of the dimer are in black; these colors are also used in the electronic-structure plots.[14] The (projected) DOS directly shows where the *s* level and the *p* levels are energetically situated, at lower (ca. −18 eV) and higher (ca. −2 eV) energies, and the lone pairs are also found at relatively high energies (0 and −4 eV). And while the DOS shows mixing between 2*s* and 2*p*, both for the σ bond and the lone pairs, it does not give **any** bonding information since – you guessed it already – the phases are

---

14 The professional quantum chemists will recognize that I am talking about (localized) bonds and electron pairs and relate them to canonic orbitals that have been plotted in Figure 3.5; while the two terms or concepts are not identical, they are close enough for a consistent interpretation.

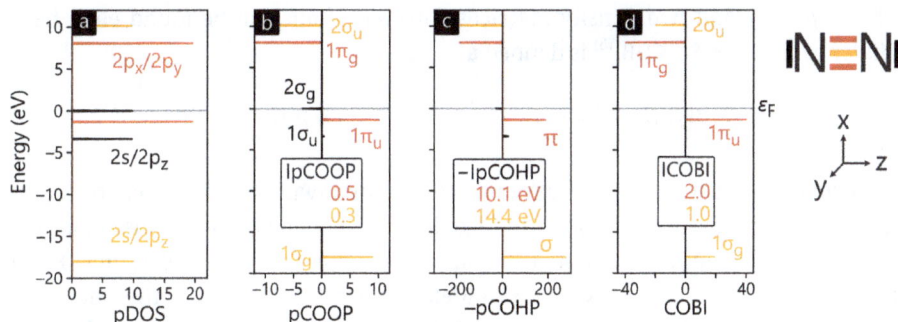

Figure 3.5: Electronic structure of the N_2 molecule (and its N≡N triple bond) as given by projected (a) density of states, (b) COOP, (c) COHP, and (d) COBI analysis calculated using LOBSTER.

missing. That is given, however, by the (projected) COOP: at lowest energy, the integrated COOP is 0.3 for the σ bond but 0.5 for the π bond. The lone pairs are visible, too, by miniscule (hence, irrelevant) bonding and antibonding levels in black; for good reasons they look as being "alone," devoid of (much) interaction. In the (projected) COHP, we see that the σ bond contributes to the band-structure energy by about 14.4 eV, and the π bonds are slightly weaker, about 10.1 eV; the COHP shows the lone pairs, too, but even weaker than in the COOP, so the two lone pairs do not lower the energy decisively. Finally, the (projected) COBI directly indicates an ICOBI (= bond order = integrated COBI up to the Fermi level) of 1.0 for the σ single bond and of 2.0 for the π double bond. Nonbonding lone pairs are totally invisible, as expected, and their bond order is zero (Müller et al., 2021). While COBI is important for molecules already, it will turn out to be even more important for solids.

Let us stick to molecules for the moment, however, and perform the almost trivial step from N_2 to O_2: the MO diagram stays about the same but one simply adds two more electrons that enter the O–O antibonding $1\pi_g$ MOs (see COOP, COHP, and COBI), thereby lowering the bond order from 3 to 2, hence, the O=O double bond. Because the $1\pi_g$ is doubly degenerate, each of the two electron takes its "own" π orbital to allow for maximum electron–electron distance, and the exchange hole generates the triplet ground state of $^3O_2$; we are constantly inhaling magnetic molecules. Fortunately enough, electronic correlation weakens the diradical character, so $^3O_2$ is chemically less aggressive than anticipated (Borden et al., 2017).

And there is even more information to extract from the COBI idea. Molecular chemistry knows plenty of examples where bonding interactions occur between more than two atoms (sometimes called two "centers"), usually a consequence of an unfitting electron count of the molecule, either too small or too large for an electron octet (see Appendix A). Because of that, one can formulate a three-center COBI, a four-center COBI, and a generalized n-center COBI. For reasons of simplicity, the three-center COBI (as defined for the crystalline case) is depicted, involving three atomic orbitals μ, v, χ and two density matrices. Further details can be found elsewhere (Müller et al., 2021). So, $\mathrm{COBI}^{(3)}$ is defined as

$$\mathrm{COBI}^{(3)}_{\mu v \chi}(E) = P_{\mu v} \cdot P_{v \chi} \cdot \sum_j w_k \mathrm{Re}\left(c^*_{\chi, jk} c_{\mu, jk}\right) \delta(\varepsilon_j(\mathbf{k}) - E) \tag{3.15}$$

Averaging cyclic permutations guarantees energetic invariance. In order to stay as simple as possible, we just look at the iconic electron-deficient molecule, diborane (B_2H_6), which cannot complete its octet since there are not enough electrons. For a thinkable BH_3 monomer, only $3 + 3 \times 1 = 6$ electrons are available, so the unit must dimerize to B_2H_6 and do something weird. The B_2H_6 molecular structure is best described by two BH_4 tetrahedra sharing an edge such that two "bridging" H atoms connect to two boron atoms at the same time. A sketch of the structure, electronic structure, and the chemical bonding is provided in Figure 3.6.

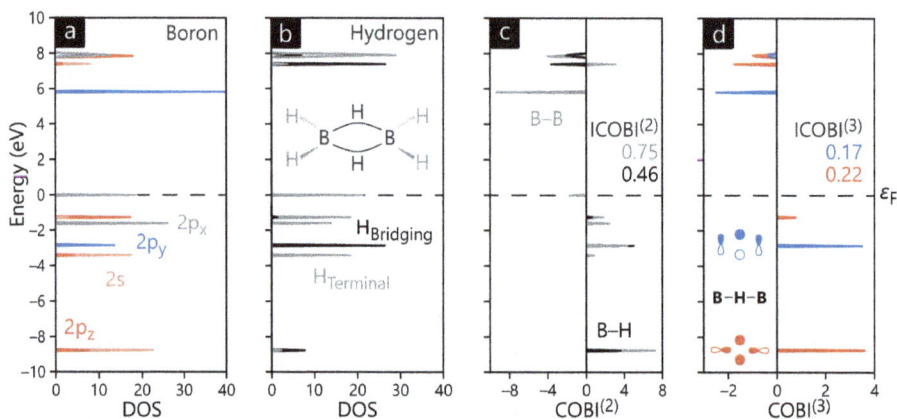

Figure 3.6: Electronic structure of diborane, B_2H_6, as seen from the projected density of states of (a) B and (b) H atoms, as well as (c) two-center and (d) three-center COBI analysis of the B–B, B–H, and B–H–B bonds calculated using LOBSTER.

The boron and hydrogen DOS let us identify at which energies these atoms, in particular H of which there are two types, contribute to the electronic structure. The two-center $\mathrm{COBI}^{(2)}$ covers B–B as well as B–H interactions, whereas the **three-center** $\mathrm{COBI}^{(3)}$ looks at the B–H–B **three-center** two-electron bond. Upon comparing DOS and COBI including the color coding, the origin of the three-center interaction is obvious: in the $\mathrm{COBI}^{(3)}$, there are occupied levels clearly attributable to the b_{1u} MO formed by the B $2s/2p_z$ orbitals and the H $1s$ orbital (shown in red), and there is also an a_g MO resulting from the B $2p_y$ orbitals and H $1s$ orbital (shown in blue). Boron's $2p_x$ orbital as well as the terminal H atoms are not involved in the three-center two-electron bonds; they mix into the B–H two-center bonds (Nelson et al., 2023). Orbital interactions are truly essential for understanding molecules (Albright et al., 2013), as alluded to before.

And there is an interesting link between the B–B two-center bond and the B–H–B three-center bond because the main COBI$^{(2)}$ peaks are positioned exactly where the COBI$^{(3)}$ peaks are. That being said, the B atoms in B_2H_6 are rather indirectly bonded via the B–H–B bond, in particular because direct bonding via $p_x–p_x$ or $p_z–p_z$ is negligible, in perfect agreement with early quantum-chemical findings (Longuet-Higgins, 1949; Lipscomb, 1966; Lipscomb, 1977). Multicenter bonding is likewise important in crystalline solids, not only for metals, as explained later.

Even when methods such as COHP and COBI are not explicitly needed, the ability to express the entire electronic structure visible in the band structure by its orbital character can be helpful, for example, when we want to know which orbital contributes to one particular band and at which k point, not averaged like in the DOS. A good example would be the two archetype polymorphs of carbon, graphite and diamond; their band structures and DOS are depicted in Figure 3.7, showing so-called **fatband** plots.

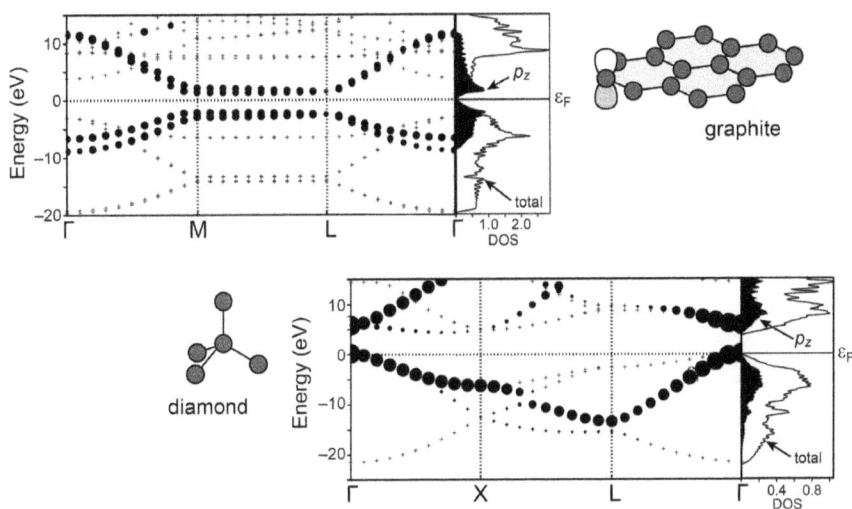

Figure 3.7: Graphite (top) and diamond (bottom) fatband analysis as regards the participation of the p_z atomic orbital on carbon.

The utility of the fatband technique is obvious from the contribution of the out-of-plane C $2p_z$ orbital of graphite (top), to be compared with the contribution of the same orbital in diamond (bottom). While simple crosses indicate the course of the energy eigenvalues through k space, the size of the superimposed circles (or intensity of color if colors were available) shows the amount of $2p_z$ participation with arbitrary scaling. And all that is mirrored in the neighboring DOS because it is just the inverse slope of the bands. In diamond, $2p_z$ is well distributed over the entire energy range, so it contributes everywhere. In graphite, however, the $2p_z$ forms the characteristic π-bonding system located around the Fermi edge, well separated from the rest, with important consequences for the physics of that allotrope (Esser et al., 2015).

Another useful utility, now explicitly touching upon the bonding characters, is given by k-dependent COHP, most conveniently illustrated from the example of thallium fluoride, TlF, a solid-state material violating the octet rule (Appendix A). Its orthorhombic crystal structure is unusual, in particular because one can imagine more symmetric structures such as [NaCl] or [CsCl] but they are not adopted. Figure 3.8 depicts both the crystal structure as well as the band structure in which the bands have been colored according to their bonding or antibonding proclivity, measured by COHP with respect to this and that orbital interaction.

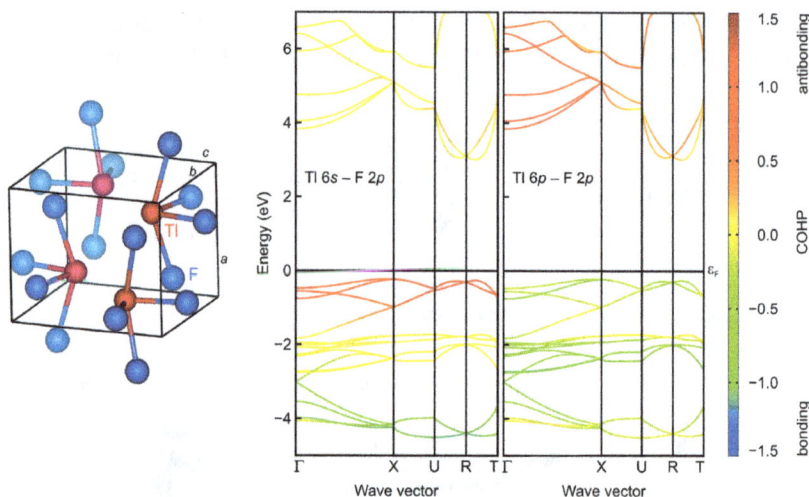

Figure 3.8: The crystal structure of TlF (left) and its band structure (right) colored according to the bonding/antibonding character of Tl 6s–F 2p and Tl 6p–F 2p orbital interactions.

There are two band structures identical in energies but differing in bonding characters. We first look at the Tl 6s–F 2p (left) and then at Tl 6p–F 2p (right) interactions. Clearly, the 6s–2p interactions are both bonding and antibonding in the valence region, but they are **massively** antibonding close to the Fermi level, so the inert pair of the monovalent Tl$^+$ atom[15] with an electron configuration of 6s^2 does engage in the highest occupied levels and is not so inert after all. To answer the question of higher symmetric structures and why the orthorhombic structure is still taken, there exist bonding interactions between Tl 6p and F 2p orbitals spreading through the valence bands, seemingly countering electrostatic interactions but only possible in the less symmetric structure which allows for such 6p–2p overlap. Such a k-dependent COHP technique is useful

---

15 Writing Tl(I) instead of Tl$^+$ is also fine.

when atoms are to be substituted by other atoms, answering – ahead of synthesis – how the bonding will change in reciprocal space (Nelson et al., 2020).

Although the interested reader may appreciate the ability to express the delocalized electronic structure in terms of atomic orbitals, thereby getting full access to questions of chemical bonding, a simple question remains: how about the density? After all, if the electron density $\rho(r)$ holds everything, at least that is the foundation of DFT, shouldn't it be possible to extract chemical bonding just from the density, in particular for solids? To answer that question practically, let us take a simple example from the solid state, the metallic elements of the main quantum number $n = 5$ utilizing $5s$ and $4d$ orbitals, so we are talking about the entire series Rb, Sr, Y, Zr, Nb, Mo, Tc, Ru, Rh, Pd, Ag, and Cd. This series comes in handy, in particular because spin polarization is not involved and, despite structural changes from *bcc* (Rb, Nb, Mo), *fcc* (Sr, Rh, Pd, Ag), and *hcp* (Y, Zr, Tc, Ru, Cd), the energetic influence of the packing (moment analysis) is rather small (Lee & Fredrickson, 2017).

For this metal series, their electron calculus will start electron poor, $5s^1 4d^0$ or better $(5s4d)^1$, and end up electron rich, $5s^2 4d^{10}$, so we expect a smooth gradual change of something between "metallic" and "covalent" from the outset. Likewise, one would love to know why the atomization enthalpy H_{at} of rubidium is 8.9 times **smaller** than the one of niobium (with the same *bcc* structure), so Nb must be more strongly bonded than Rb by almost one order of magnitude. Likewise strontium is 3.4 times **less strongly bonded** than rhodium (both *fcc*), and zirconium is 5.5 times **more strongly bonded** than cadmium (both *hcp*) based on H_{at}. What is the reason? And does it materialize from the density? Figure 3.9 provides the density information $\rho(r)$, showing the most densely packed layers, that is, coinciding with (110) in *bcc*, (111) in *fcc*, and (001) in *hcp*.

Despite the obvious shell structure close to the atomic nuclei (with rapid color changes), there isn't really much detail in the density **between** the atoms where the chemical bonding is expected to take place. In fact, the in-between density for Nb is marginally higher than in Rb, so one may **guess** that the bonding will be slightly stronger in the first place. Similar guesses can be made for the Sr–Rh and the Zr–Cd cases, but these are guesses, simply because the **phases** are missing in the density, just like in the DOS. Hence, no information whatsoever about bonding or antibonding is provided since this precious information gets lost when the wave function is integrated to yield the density:

$$\rho = \int \psi^* \psi \, d\tau \qquad (3.16)$$

So, the wave function determines the density unambiguously, but the reverse is not true.[16] Hence, chemical bonding – which rests on the wave function – is difficult, if not impossible to quantify from the density **alone**, similar to a well-known problem

---

16 This is the reason why (molecular) quantum chemists value the wave function over the density, in particular because the wave function is experimentally accessible, too (Schwarz, 2006).

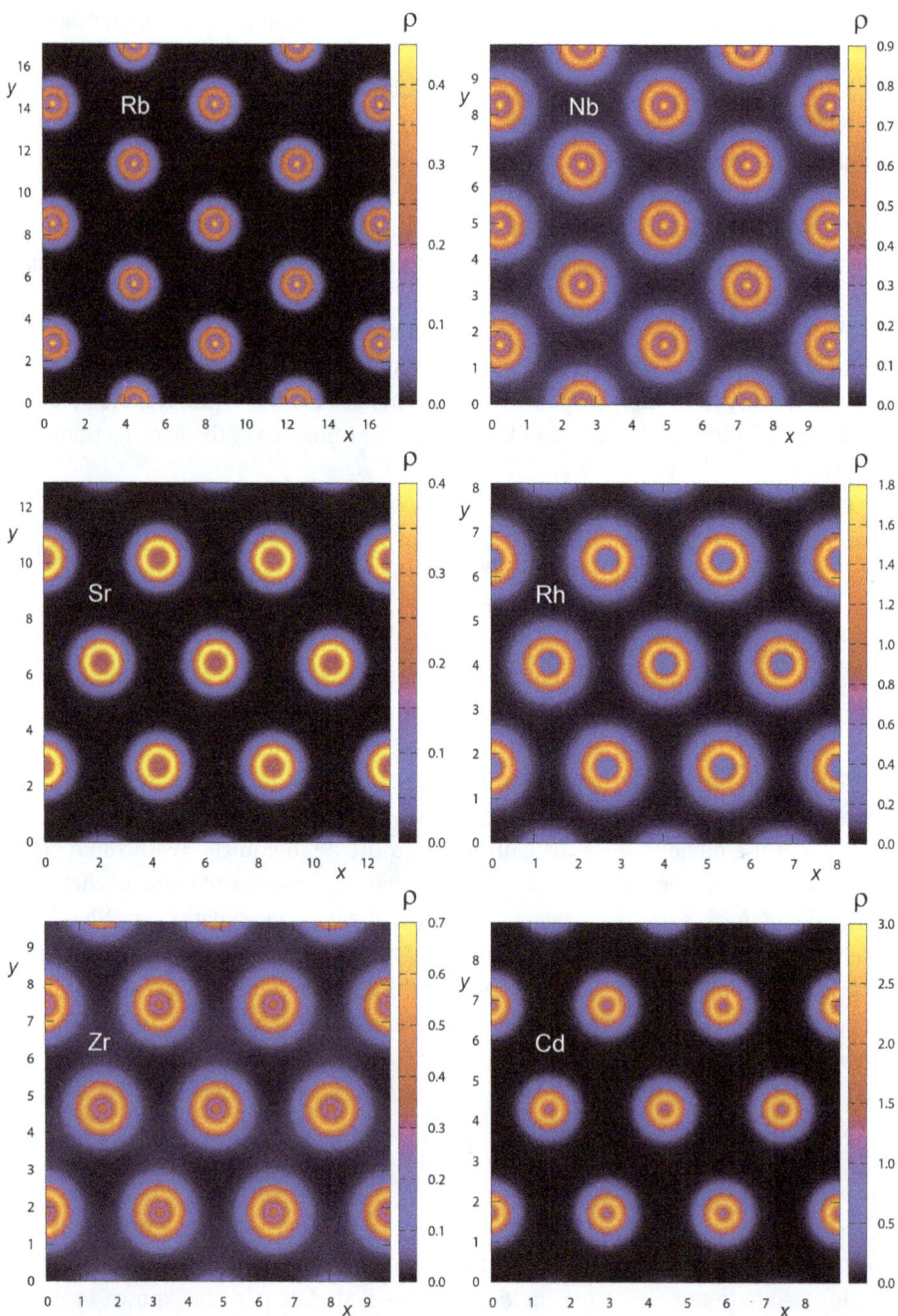

Figure 3.9: Two-dimensional electron densities ρ given in e (Bohr radius)$^{-3}$ within dense planes of elemental metal structures from the fifth period, *bcc*-like on top (Rb and Nb), *fcc*-like in the middle (Sr and Rh), and *hcp*-like at the bottom (Zr and Cd), with all distances x and y given in Å.

encountered in crystallography.[17] And still, bonding (in metals) is sometimes discussed based on the density only, for whatever reason.

Being aware of the density problem, other theoretical methods have been developed to quantify the **localization** of electrons, a key ingredient of covalency where electrons are localized and shared between atoms. This is the intention of the electron localization function (ELF), a standard real-space visualization tool (Becke & Edgecombe, 1990) encountered in many publications. The ELF definition is intimately connected with the pair probability of like spins (and the kinetic-energy density of the wave function), and while the tricky details may be skipped at this point (Dronskowski, 2005), ELF is designed in such a way that it is numerically bound between 0 (fully delocalized) and 1 (fully localized electron). An insightful analysis (Burdett & McCormick, 1998) shows that the ELF behavior is basically given by the nodal properties of the system's occupied orbitals (i.e., the kinetic-energy density, as said before).

In addition to having witnessed the electron density, let us now study the localization of the electrons, so we repeat the aforementioned visual analysis of those six metal structures but now plotting ELF instead of $\rho(\mathbf{r})$, depicted in Figure 3.10.

Admittedly, Figure 3.10 is nicely colorful and aesthetically pleasing but, despite a stronger nodal structure than in $\rho(\mathbf{r})$, there isn't much additional information, actually less. The in-between localization of electrons in the bonding regimes in Rb and Nb is practically the same, both for nearest and second-nearest neighbors. For Sr, the electron localization is even **larger** in the three-atom regime than for Rh, although rhodium is much stronger bonded; the two-center in-between localization for Sr and Rh is the same. Likewise, Zr and Cd are almost indistinguishable in terms of ELF in the in-between bonding regimes. It is hard to extract any meaningful information for the amount of bonding from the ELF which also lacks the phases. Evidently, ELF does not correlate at all with the bonding energy.

Upon taking into account the phase information provided by the wave function, for example, by including the density matrix $P_{\mu\nu}$ needed to calculate the two-center COBI, the bonding mechanism is easy to explain, almost trivial, as shown in Figure 3.11.

As the electrons are filled into the orbitals of the $5s4d$ metals, the atomization enthalpy (left, in red) grows, so bonding strengthens until approximately half filling (around Nb to Tc) is achieved; additional electrons then fill antibonding levels such that H_{at} diminishes (Cd). And the same course is mirrored semiquantitatively from the integrated COBI curve (in blue), which shows the bond order of a single atom–atom contact; for the example of Rb, ICOBI = 0.12, which corresponds to a total bond order of 0.98 of

---

17 Even though the observable real-valued intensity I of an X-ray diffraction experiment **appears to contain** the entire information, it does **not** do so because the crystal structure is unknown **unless** the phase problem is solved, thereby yielding the atomic positions and, hence, the (generally complex) structure factor F which is then squared to yield a theoretical I, to be compared with experiment for validating the structural model. The mathematical analogy between quantum mechanics and crystallography is compelling.

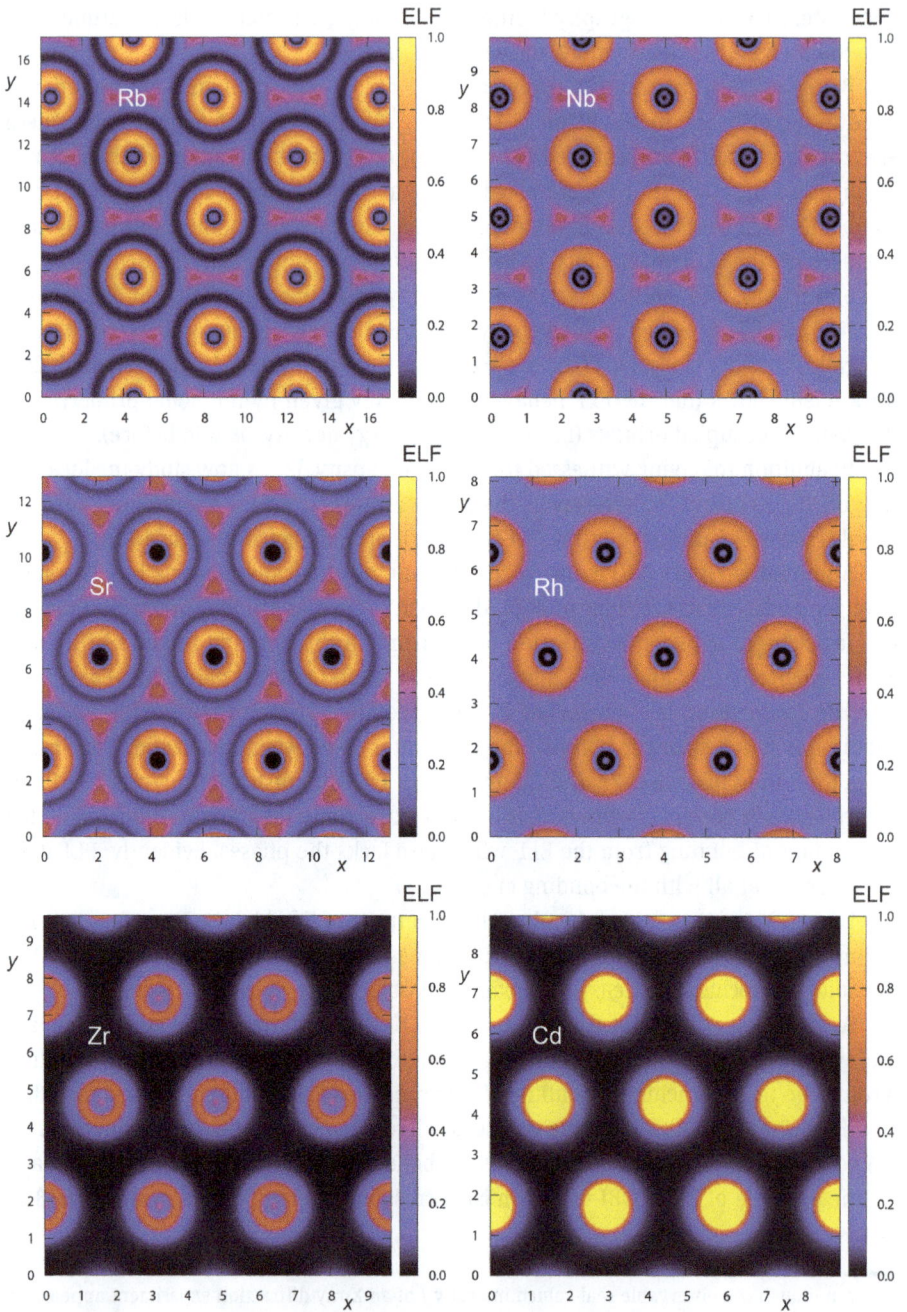

Figure 3.10: As before but now showing the electron localization function (ELF).

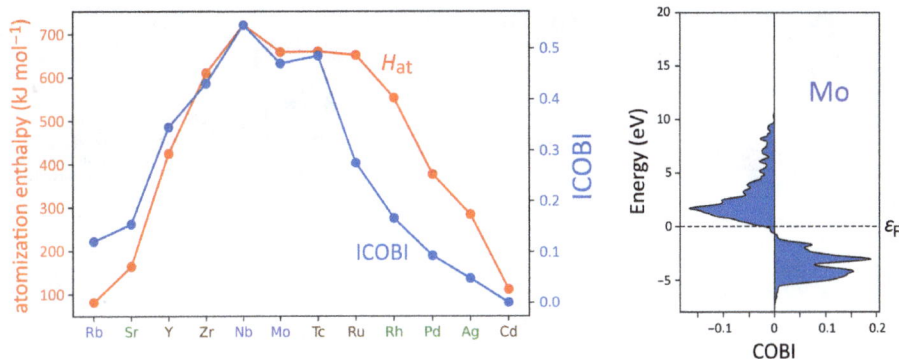

Figure 3.11: Atomization enthalpy (red) and two-center ICOBI (blue) of various metals of the $5s4d$ row (left) and their relationship to the energy-resolved COBI plot for the example of Mo (right).

Rb to its eight nearest neighbors. The COBI course itself can be understood from the energy-resolved COBI plot of molybdenum (Figure 3.11, right), a fine example of the rigid-band model. For low electron counts, only bonding levels are occupied, whereas Mo itself already has a few antibonding levels filled; hence, it is less strongly bonded than Nb. Electron-richer atoms than Mo are less strongly bonded because even more antibonding levels are filled up. Please note that the curves of H_{at} and ICOBI differ more for large electron counts, here many-body effects that do not fully translate into the one-electron density matrix become apparent, and there are obvious limits to the simple rigid-band model.

These illustrations of solid-state bonding and its analysis based on atomic orbitals and their phases may suffice for the moment being, so let us get slightly more technical, so to speak. We already alluded to the fact that the unitary transformation between the totally delocalized representation (plane waves) and the totally localized representation (atomic orbitals) can be performed automatically, given a good electronic-structure calculation to be built upon and a reliable atomic-orbital basis set. This very task is carried out by the LOBSTER code, freely available for academic users at www.cohp.de, ready for download. First, one needs a fine DFT calculation, performed by either VASP, Quantum ESPRESSO, or ABINIT, all three of them will do nicely. LOBSTER then reads the entire plane-wave DFT data and makes some intelligent guesses which kind of local basis set should be used, and it even gives a few useful tips to the user which orbitals are needed. The LOBSTER flowchart (Nelson et al., 2020) is presented in Figure 3.12.

LOBSTER can be run even without an input file but personally (yes, I identify as Mr. Old-fashioned) I would recommend to have a lobsterin file in which the orbital basis is directly specified by the user (= you) through the basisSet command. You may also specify the basis set individually per atom using the basisFunctions command. And then the entire analysis may start, for example, by calling the powerful

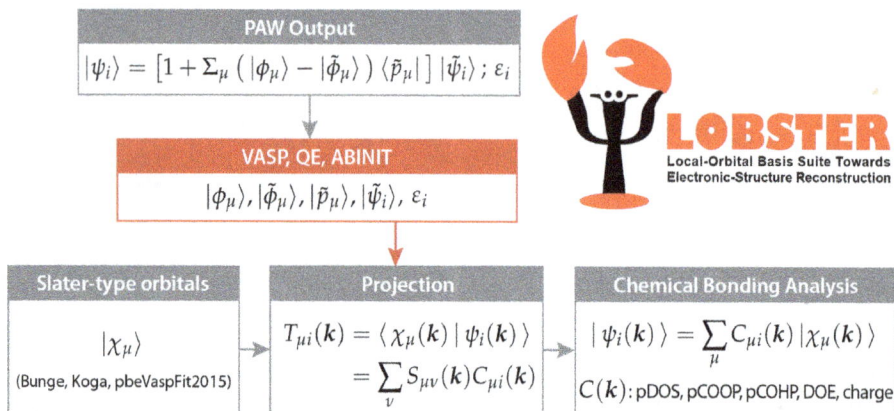

Figure 3.12: Flowchart of the LOBSTER computer program.

cohpBetween by specifying which atom pair is to be looked at. For a finer analysis, the COHP calculation can be carried out orbitalWise or may involve all (on-site and off-site) elements of the density matrix, so DensityOfEnergy may be another option. By default, Mulliken and Löwdin population analyses are done automatically, and this also includes the calculation of (numerically reliable) Madelung energies.

For the expert, one may use basisRotation to align certain orbitals, one may use the createFatband command to fatten bands according to their orbital character, and one may color the bands according to their bonding proclivity (the kSpaceCOHP command). You may also be interested in using LSODOS to arrive at a DOS that has profited from Löwdin's symmetrical orthonormalization such as to cure a deficiency of PAW theory. And you may call the mighty cobiBetween command to analyze various two-center, three-center, four-center, etc. COBI, yielding bond orders. Of course, all the wave functions can be read out and plotted, according to your choice. You guessed it already, there are many options, and the program is extremely powerful. Go and download LOBSTER!

4 Seven ~~Samurai~~ material classes and their chemical bondings

The careful reader will recall that there are three strong and, at the same time, fundamental chemical-bonding types: covalency, metallicity, and ionicity. The Almighty not only invented them but also superposed a somewhat weaker but nonetheless truly fundamental bonding type, dispersion interactions (stemming from electronic correlation), on top of the first three. For reasons of sheer convenience, a fifth bonding type – **not** a bonding type of its own – has been formulated by lazy humans: hydrogen bonding, an interesting mélange of covalency, ionicity, and dispersion interactions. In principle, all these may come together, and they do exactly that to generate the entire universe of chemical bonding in another universe of chemical compounds. How should one navigate and not get lost?

If this were a book about chemical bonding in organic matter, there would be a simple solution. To an astonishingly good approximation, the entire organic chemistry (simply speaking, molecular compounds of the first **long** period, also called second period) is held together by covalency. There is a very weak influence of ionicity, mostly ignored; the molecules "touch" each other by dispersion forces, also to be ignored (except in thermodynamics), but a little hydrogen bonding is often encountered, say, in the solid-state chemistry of organic matter bravely dubbed "supramolecular," a funny term worth a Nobel Prize. Crystals of organic (and a few inorganic) matter sometimes show different structures than the isolated molecules themselves, and this is the reason for the naming, also slightly simplified.

But this book is mostly about inorganic solids, solid-state materials, so to speak, and here the problems begin. Different scientific communities have found different ways to organize exactly the same chemical species. Also, the entirety of solid-state chemistry covers all elements of the periodic table, so even the literature is exceedingly rich or "diverse" (Dronskowski et al., 2017). In addition to that, it is rather difficult to find competent treatises of solid-state chemical bonding from most quantum-chemistry textbooks, the reason being that the latter mostly focus on molecules. An exception is made as regards an early jewel of chemical theory (Hoffmann, 1988), partially going back to the ingenuity of the author (Seeman, 2022a; Seeman, 2022b) and, also, the flexibility of the underlying (fairly simple) semiempirical theory to be used for **any** matter, molecules, surfaces, and solids, too. Likewise, there are other excellent books (Harrison, 1980; Burdett, 1995; Burdett, 1997), and also those targeting solid-state chemical bonding from the density-functional theory (DFT) perspective (Dronskowski, 2005). Two recent reviews (Miller et al., 2017; Jones, 2022) put things into perspective by focusing on chem-

---

Note: Sorry, I had to write that because this book was composed in Japan. And if you do not know Akira Kurosawa's movie, you better watch it.

ical interpretations but also by considering the historical aspects of solid-state bonding theory.

For the moment, I will take a different, more heuristic route which is easily explained. For very good reasons, inorganic **chemists** organize (solid-state) compounds according to the participating elements because the periodic table of the elements is decisive. So, we could equally group all the materials into halides, oxides, nitrides, arsenides, ferrates, phosphates, etc., a fine way to do it, typically found in excellent inorganic textbooks (Holleman et al., 2007; Greenwood & Earnshaw, 1997).The chemical bonding in a refractory oxide such as CaO and a molecular oxide such as Mn_2O_7 could not be more different, however, so this guiding principle will **not** serve us well.

Another way to sort solid-state materials would be given by the crystal structures (done by our **crystallographer** friends) or by even simpler stoichiometry-based principles (by people from **computational materials science**). In the first case, we would put all compounds crystallizing in the [CsCl] structure type with space group $Pm\bar{3}m$ together but the chemical bonding in, say, TlBr and AlFe truly differs; this is not a good choice. In the second case, all AB_2-type compounds such SiO_2 and FeS_2 are arranged together, probably a good choice for a computer program but light years away from chemical understanding; actually, this is clearly wrong. The computer does not care but we do.

And then one may group materials according to their physical properties, typically carried out by the **physicists**, and it is a reasonable choice, not bad at all to begin with. Thus, we may divide insulators, semiconductors, and metals, for example, but then there will be at least two problems. First, conductivity does not necessarily imply the underlying bonding type (if you compare the two metals graphite and sodium) and, second, there are complex compounds that incorporate various kinds of bonding at the same time. Yes, different bondings may coexist in the same material.

Because of these many difficulties, I decided to group little more than a handful of important and (according to my nonobjective estimate) representative compounds into seven classes such that there is a **common chemical-bonding principle** in each of the seven groups, at least approximately. The magnificent seven[1] start with simple compounds, either mostly covalent or mostly ionic in character, which fulfill the very important octet rule of general chemistry. Second, reiterating on mostly ionic materials, there is a group of solid-state matter in which more than one anion defines the chemical bonding (mixed-anion compounds), and given so-called complex anions the systems further complicate. Such systems are often unknown outside chemistry, so I am also trying to teach the non-chemists a little chemistry, and I hope they will appreciate that. Third, there are those compounds whose fine bonding mixture between covalency and ionicity defines the chemical reactivity, and it may rapidly change as a function of composition.

---

1 There is a similar, yet different book that I highly recommend, a little dated, but still a pleasure to read (Moore, 1967).

Here I thought that this group – for which covalency and ionicity effectively compete – would also be a good group, in particular because we are now talking about the very timely group of battery materials. Fourth, it seems highly fitting to talk about solids that are made up from molecular units, and how these molecular compounds are held together when forming crystals. We will witness somewhat weaker forces such as the aforementioned hydrogen or other secondary bondings, as they are sometimes called. As regards group number five, we will dive into intermetallics which, depending on the electronegativities (ENs), show up as semiconductors (yes, that is true) or also metals. Metallic bonding is a fascinating subject; it comes in many varieties and eventually leads to important construction materials such as steels. Under high pressure, the simple rules of chemical bonding may need modification here and there, at least there are new marvels not encountered under standard conditions, so there is group number six comprising so-called high-pressure phases as another chemical-bonding group, not only relevant to geoscience. Eventually, there are those solid-state phases where covalent (but not metallic) interactions involve more than two neighboring atoms, for an electron-count reason, a phenomenon seen before and actually well-known for molecules but less known (or almost unknown) for solids. This group involves multicenter bonding, so it is important to really cover it and define chemical-bonding group number seven.

That being said, please get ready for an exciting tour through solid-state chemical bonding in all its varieties; it will be fun.

4.1 Simple systems in harmony with the octet rule

Let us start with the first chapter dealing with supposedly "simple" systems, at least from the perspective of general chemistry and the associated electron count. As we are also moving into the crystalline world and need to deal with periodic, that is, translationally invariant systems, a first *Gedankenexperiment* seems to be fitting.

As covered in Chapter 3 already, two H atoms readily form an H–H dimer such as to acquire the He electronic configuration $1s^2$ (or [He] in short) for each of the H atoms. We are effectively counting the two shared electrons **twice**, both for the left and the right atoms, but since nature does so, too, this will not bother us any longer; we are on the safe side. Because of two interfering atomic orbitals, there result two molecular orbitals of which one has a bonding character, and it gets "filled" by two electrons, so this yields a stable molecule with an H–H distance of 0.74 Å (see Chapter 3). Hence, one faces an H–H dissociation energy of about 431 kJ mol^{-1}, a truly substantial amount for the prototype covalent bond. Let us then think of making many H_2 molecules to **polymerize** such as to form a one-dimensional crystal[2] as depicted in Figure 4.1, the aforementioned *Gedankenexperiment*:

---

2 As alluded to already, one-dimensional crystals do not exist, they are tools to ease understanding.

···· H ···· H ···· H ···· H ···· H ···· H ···· H ···· H ···· H ···· H ····

Figure 4.1: Schematic sketch of a one-dimensional hydrogen chain.

Given a reasonable H–H distance of about 1 Å (a little wider than in the molecule since each H atom now needs to bond to left **and** right at the same time, despite the same electron count, just one e per H atom), the simplest band structure of that system is easily calculated from semiempirical extended Hückel theory, simply by having used Bloch's theorem, and it is depicted in Figure 4.2.

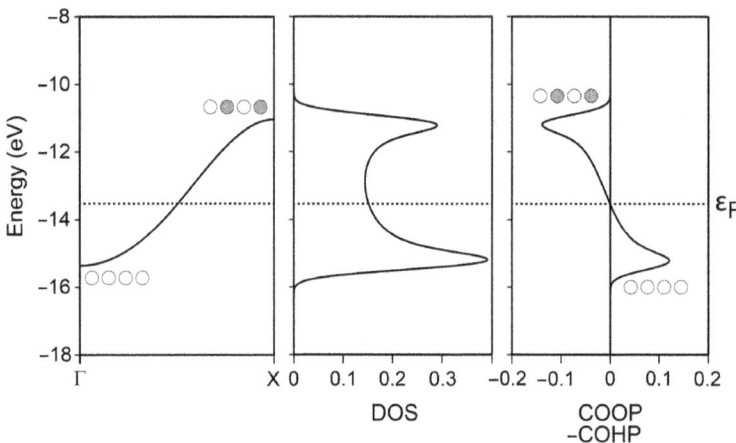

Figure 4.2: Qualitative band structure (left), density of states (DOS, center), crystal orbital overlap population (COOP, right), and crystal orbital Hamilton population (COHP, also right) of the equidistant hydrogen chain spaced at H–H = 1.0 Å based on an electronic-structure calculation using extended Hückel theory.

There is an electronic band energetically moving up when going from the zone center **Γ** to the zone edge **X** of the one-dimensional Brillouin zone, and we are witnessing the crystallographic equivalent to the molecular-orbital diagram of H_2 shown in Chapter 3. The course of the band (either up or down) depends on the orbital topology, and the width of the band (its so-called dispersion) grows with decreasing interatomic distances; it is that simple. The density-of-states (DOS) diagram just serves as a level counter because it averages over the entire Brillouin zone, and for the actual electron count (1 e per H atom and band), the DOS would be half-filled, and one-dimensional crystalline H would be a metal, with the Fermi level around −13.5 eV, tentatively sketched with a dashed line. For this very electron count, both crystal orbital overlap population (COOP) and crystal orbital Hamilton population (COHP) indicate that all the bonding levels are filled; hence, the crossover from bonding to antibonding essentially **defines** the position of the Fermi level. So, there is nothing to worry about – but only at first sight. A deeper analysis (Hoffmann, 1988; Dronskowski, 2005) shows, however, that such a half-filled one-dimensional system is subject to structural deformation, as proposed by Peierls[3] (1955), and the one-dimensional system will decompose into H_2 molecules, as guessed by any trained theoretical chemist. It is fun to study how the distortion happens (Hoffmann, 1988; Dronskowski, 2005) but we can safely skip it for the rest of this chapter.

Now that the principles of band structures, DOS, and chemical-bonding indicators have been illustrated, once again, let us move toward a three-dimensional real-world (not model) system, and let us also utilize a better theory that properly deals with electronic exchange and correlation, that is, DFT. Figure 4.3 shows the results of a corresponding DFT calculation on crystalline silicon in the diamond structure (with each Si atom tetrahedrally bonded to four nearest-neighbor Si atoms) using a relatively simple exchange-correlation functional, the GGA. Indeed, this is the silicon equivalent to the carbon system (also crystallizing in the diamond structure) already shown in Figure 3.3.

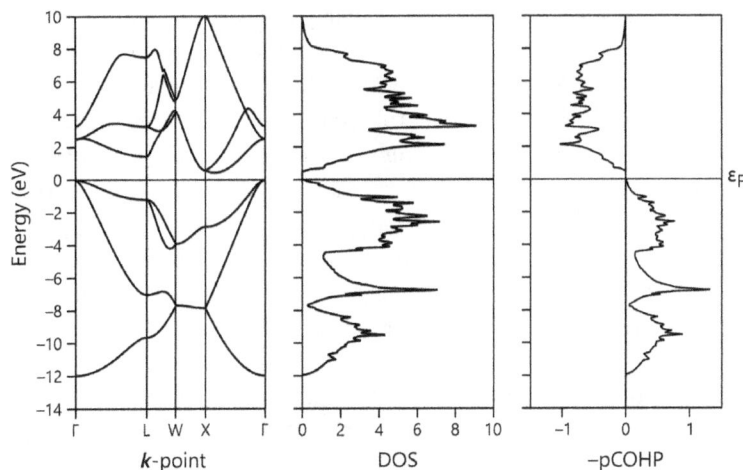

Figure 4.3: Band structure (left), density of states (center), and projected crystal orbital Hamilton population (projected COHP, right) of crystalline silicon in the diamond structure based on a GGA electronic-structure calculation.

In contrast to the preceding example, in real-world silicon, there are bands running up and down while walking through reciprocal space, but we need to memorize that **two** kinds of atomic orbitals (3s and 3p) are involved, and their topologies differ from each other, depending on the course taken within the reciprocal space. We are also

---

3 The Peierls distortion (**not** Peierl's distortion) is named after Rudolf Peierls (1907–1995), a German-born British theoretical physicist who proposed the effect. It is quite enjoyable to study an account by Peierls (1993) about the early days of solid-state quantum mechanics.

witnessing that the top of the valence bands (at 0 eV) is **not** directly below the bottom of the conduction bands, so elemental silicon is an **indirect** semiconductor, as known from solid-state physics. Clearly, the DOS figure yields a bandgap of less than 1 eV, a bit too narrow compared to reality (ca. 1.1 eV) but the rather inexpensive exchange-correlation functional has not been designed for such property. And if we compare Figure 4.3 with Figure 3.3 (diamond case), it is quite obvious that the dispersion in the valence and also conduction bands is smaller, and even the silicon DOS is less smooth than the carbon DOS. In fact, $3s$ and $3p$ (Si) do not mix as effectively as $2s$ and $2p$ (C) because Si is not a member of the first **long** period (Appendix E), sometimes also called second period.

The COHP (note that we are now exploiting variational DFT for which COHP is the better choice) curve also shows what is to be expected: the entire valence band of Si is bonding, just like in diamond, and the conduction band is entirely antibonding, **hence** (as the chemists would phrase it) the crystal structure is perfectly stable for a Si–Si inter-atomic distance of 2.35 Å, about twice the tabulated covalent radius of silicon (1.18 Å). Any substitution of Si ($3s^2\,3p^2$) by an electron-richer main-group atom such as P ($3s^2\,3p^3$) would immediately weaken this structure's bonding because the surplus electron had to go into the antibonding conduction bands. And please note that elemental silicon, with an electron count of 4, perfectly fulfills the octet rule since each single bond to the four nearest-neighbor silicon atoms carries **two** electrons, so each Si atom is surrounded by **eight** valence electrons and silicon's electronic configuration corresponds to the [Ar] con-figuration, a stable electron **octet** of a noble-gas atom, at least formally.[4] It may sound rather trivial for a trained chemist but let us repeat that one more time: while each iso-lated Si atom has four valence electrons, its bonding situation with four tetrahedrally bonded Si atoms makes it reach an **effective electron count** of eight, the noble-gas elec-tron count, simply speaking.

Silicon's atomic electron count could also be reached if one would replace one Si atom ($3s^2\,3p^2$) by one Al atom ($3s^2\,3p^1$) and another one by P ($3s^2\,3p^3$) because the arith-metic average for the AlP combination (3 + 5 = 8) is also four. So, instead of having two atoms from main-group IV,[5] we might want to replace them by one atom from main-group III and another one from main-group V, very simple indeed. Because we would like to deal with real-world examples that also find application in modern technology,

---

4 That being said, please note that the heavier noble gases are reactive and form covalently bonded molecules, so they are **violating** the noble-gas electron configuration (Hoppe, 1964), as detailed in Ap-pendix A.

5 According to IUPAC, main-group IV should better be called group 14 because they also took the $10d$ electrons of the transition metals into their consideration. IUPAC did not consider, however, to include the $14f$ electrons of the lanthanide and actinide elements, so group 14 really is group 28 in the **long** periodic system. Admittedly, the IUPAC system is handy (printable on A4 paper). For simplicity, I stick to the old – and trusted – naming system also used in semiconductor physics. Life is too short to com-plain about nomenclature. See also Appendix E.

let us calculate gallium nitride (GaN) which forms the basis of modern optoelectronics. Ga is the higher homologue of B, also from main-group III, so the aforementioned calculus must hold true as well, and it does. This, by the way, would be the right time to get yourself a periodic table of the elements,[6] one attached to the wall, the other one below your pillow (because then all the chemical wisdom may slowly diffuse into your brain overnight, I am smiling).

Let us look at the DOS figure of GaN depicted in Figure 4.4, together with those of GaP and GaAs. To perform the needed calculations, we have assumed the wurtzite structure type in which the Ga atom is also tetrahedrally coordinated by N (or P or As) atoms, and each N (or P or As) atom is also tetrahedrally coordinated by Ga atoms, the known motif from the diamond structure:

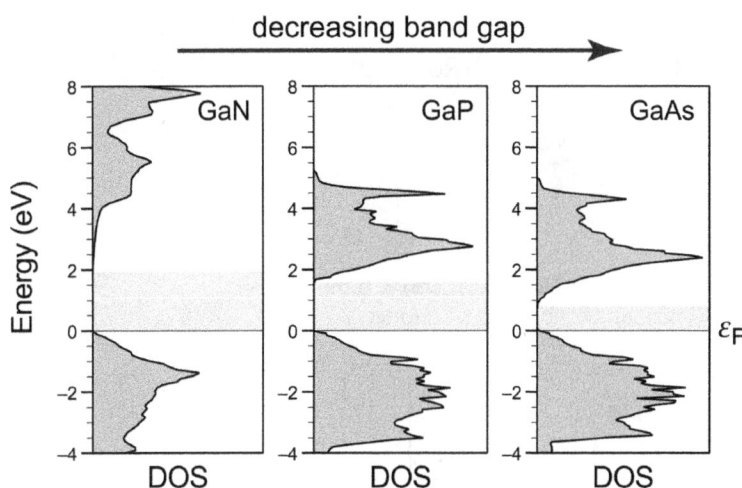

Figure 4.4: Density of states of GaN (left), GaP (center), and GaAs (right) in the wurtzite structure type based on a plane-wave PAW electronic-structure calculation.

GaN is a typical (direct) semiconductor with a theoretical bandgap of about 2 eV (in reality: 3.4 eV), and we will ignore the numerical deficiency of the exchange-correlation functional for the moment, the qualitative course serves us perfectly well. The important point to make is that the bandgap **shrinks** when the difference in (absolute) ENs (see also Appendix B) between Ga and N ($|3.2\ \text{eV} - 7.30\ \text{eV}| = 4.1\ \text{eV}$) is lowered, for example in the case of GaP ($|3.2\ \text{eV} - 5.62\ \text{eV}| = 2.42\ \text{eV}$) and even more so for the GaAs ($|3.2\ \text{eV} - 5.30\ \text{eV}| = 2.1\ \text{eV}$) case. That is to say that the more covalent these III–V semiconductors become, the smaller the resulting bandgap, either from simple DFT (2, 1.5, and 1 eV) or

---

6 Many years ago, my academic teacher Dr. Hartmut Plautz told me that without knowledge of the periodic table, inorganic chemistry would always seem like a cabinet of rarities. Boy, was he right!

from experiment (3.4, 2.3, and 1.4 eV). Reiterating, such train of thought is rather straightforward for systems that follow the octet rule (Deringer & Dronskowski, 2013).[7]

But let us now move really away from the diamond or wurtzite examples and compare them, instead, with well-known materials that do **not** share the same structure type and that strongly deviate in terms of **stoichiometry** and main groups; nonetheless, let us stick to the octet rule in such a way that each individual atom must acquire a noble-gas electronic configuration. The well-known SiO_2 α-quartz mineral and structure type (here the two most abundant elements in the Earth's crust, O and Si, get together) and the likewise well-known NaCl rock salt mineral and structure type will serve as beautiful and powerful examples. Their structures are shown in Figure 4.5, together with their energy-resolved crystal-orbital bond indices (COBI) and their integrals (ICOBI), in addition to wave function-derived Löwdin charges.

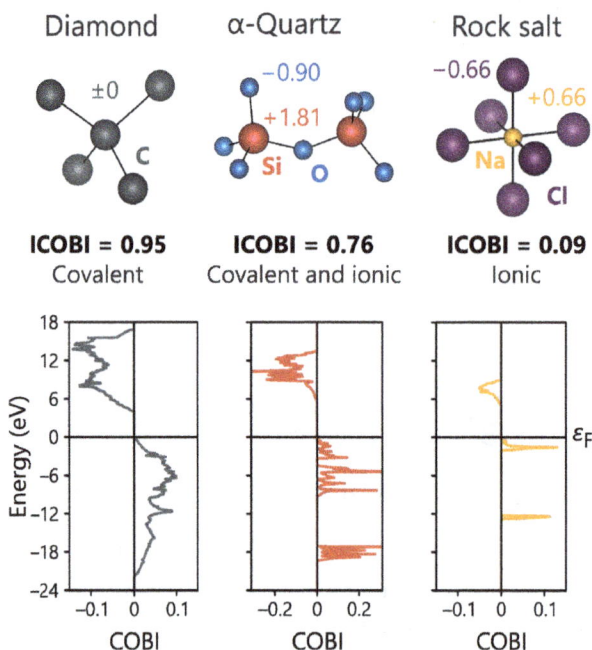

Figure 4.5: Structural and bonding motifs found in the crystal structures of diamond (left), α-quartz (center), and rock salt (right), together with integrated/energy-resolved crystal orbital bond indices (ICOBI/COBI) and wave function-derived Löwdin charges.

---

7 And while the idea works nicely for periodic solids, the situation is different for, say, diatomic molecules because the more ionic CO has a smaller HOMO–LUMO gap than the perfectly covalent N_2 (Bickelhaupt et al., 1998).

We remind ourselves that C in the C–C-bonded diamond structure fulfills the octet rule, and its electronic configuration then corresponds to [Ne]. For α-quartz, the ionic limit (boldly assuming Si^{4+} and O^{2-}, not a terribly bad choice actually) is helpful, meaning that the octet rule is also valid because both Si^{4+} and O^{2-} correspond to [Ne], just like C before. And even for NaCl in which the ionic limit (Na^+ and Cl^-) must be a very good starting point, the octet rule is fulfilled because Na^+ corresponds to [Ne], once again, whereas Cl^- resembles [Ar]. Fortunately, electronic-structure theory allows us to be much more quantitative, thanks to COBI and its integral, providing direct access to the bond order, irrespective of differences in (band) structures; hence, we do not need to plot them at all.

For diamond, the ICOBI yields a bond order of 0.95, that is, a C–C single bond with two electrons shared between neighboring C. Because there is no EN difference between all the C atoms, there cannot be any net charge transfer, and the Löwdin charge of ±0 confirms that. The energy-resolved COBI, resulting from a broad valence-band region below the Fermi level, also indicates massive dispersion that results from covalent interactions; please compare with Figure 3.3. For α-quartz, however, there is a weaker covalent bond between Si and O, with a bond order of 0.76 (so, a ¾-Si–O bond, among brothers, plus ionic attraction), and the increasing influence of ionic bonding going back to an EN difference is reflected both from the Löwdin atomic charges of +1.81 (Si) and −0.90 (O) and also from the more "spiky" course of the energy-resolved COBI. Finally, there is NaCl whose energy-resolved COBI consists of just two spikes below the Fermi level, an almost insignificantly small covalent Na–Cl interaction (ICOBI = 0.09) but, instead, strongly charged ions (±0.66) leading to a likewise strong Madelung field, stabilizing the entire collective. Remember that, even in NaCl, there is a tiny amount of covalency. Clearly, we have just witnessed a gradual change from a covalent (C) over a mixed covalent-ionic (SiO_2) to an ionic (NaCl) system, so easy to quantify using the right quantum-chemical tools (Müller et al., 2021). And one should not forget to mention that the aforementioned atomic charges as derived from the quantum-chemical wave function are **much** closer to reality than those "ideal" charges (+1 for Na, −2 for O) often mentioned in introductory solid-state chemistry textbooks; the latter are far from the truth (Appendix F).

The reciprocal relationship between covalency and difference in EN alluded to before is almost trivial to demonstrate, so let us do that for various binary compounds; in fact, it is a little easier to do that for finite objects (molecules) than for solids but the principle is the same. To do so, we would just plot the ICOBI (= bond order) against the EN difference, say, by the absolute scale (Pearson, 1988), see also Appendix B, and that is mapped in Figure 4.6.

If the difference in EN vanishes, the bond order approaches unity, and the hydrogen fluoride (HF) molecule departs somewhat from that relationship; see Figure 4.6 (left). Likewise, a very large EN difference translates into an insignificant bond order. This very relationship was already recognized by Pauling (1960), albeit using a different definition of ionicity (which, in Pauling's case, scales with the molecular dipole moment), so we also provide it in Figure 4.6 (right) for reasons of comparison, this time based on EN differences as given by Pauling. It is fun to rediscover those fundamental principles based on DFT

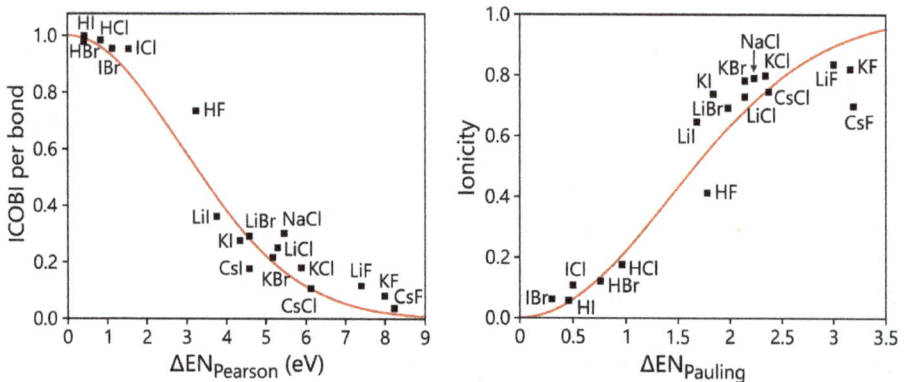

Figure 4.6: ICOBI values versus electronegativity differences (left, Pearson scale) for a number of diatomic molecules based on density-functional theory. For comparison, ionicity values as defined by Pauling versus electronegativity differences (right, Pauling scale) are also given.

calculations and how they may be analyzed by tools such as COBI (Müller et al., 2021). And one may quantify how well the historically defined ionicity (by Pauling) and an alternative ionicity definition based on Löwdin charges compare with each other. To do so, the ionicity I_{charge} of a binary solid-state phase – now we get back to crystals – composed of main-group elements (all following the octet rule) just depends on the Löwdin charge q_i and the effective valence $v_{eff,i}$ of that ion (+1 for Na, +2 for Mg, −1 for Cl, −2 for O, and so forth). The empirical formula (or handy definition) reads:

$$I_{charge} = \frac{1}{N_{Atoms}} \sum_i^{N_{Atoms}} \left(\frac{q_i}{v_{eff,i}} \right) \tag{4.1}$$

Figure 4.7 provides the proper plot, showing a good general trend but also differences, in particular for phosphides, antimonides, etc., which the quantum-chemical approach depicts as less ionic than the Pauling approach. Both ionicity definitions, however, characterize alkali halides as ionic, and CsF is the most ionic of all. For LiI, Pauling is in favor of ionicity (0.89), whereas quantum chemistry opts for less ionicity (0.55) (Nelson et al., 2023). In fact, there may be a problem as regards Pauling's definition in overestimating ionicities, suspected for a long time already (Hannay & Smyth, 1946).

For reasons of completeness, it is probably fair to say that plenty of different ionicity definitions have been developed over the historical course of – mostly – crystal chemistry, and they are all based on whatever kind of atomic charges, an almost trivial statement. And it has been shown, by means of statistical analyses, that all those atomic charges are connected with a **principal component** of ionicity accounting for more than 90% of the whole variance in the set of charges (Meister & Schwarz, 1994).

Should one stop at binaries? Of course not, ternary or quaternary or whatever phases will do just as well. For illustration, let us look at Figure 4.8 showing structure

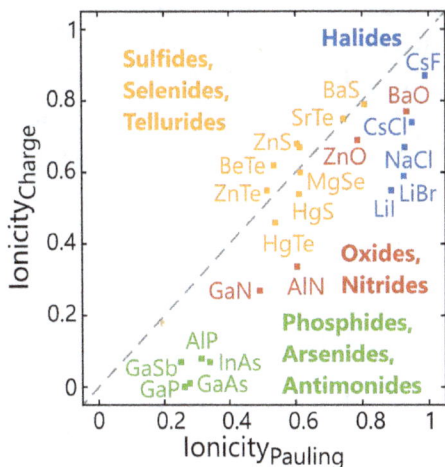

Figure 4.7: Ionicity of binary solid-state phases as derived from wave function-derived Löwdin charges against Pauling's definition of ionicity from electronegativities.

and bonding in $BaTiO_3$, an iconic ternary oxide adopting the perovskite structure type which is of paramount importance (Shimakawa, 2017) for electrical engineers; this phase and related ones are incredibly useful when used as dielectric materials (Itoh, 2017; Waser & Hoffmann-Eifert, 2017). For completeness, let us state that all the atoms involved also follow the octet rule. In the ionic limit, Ba^{2+} corresponding to [Xe], Ti^{4+} corresponds to [Ar], and O^{2-} corresponds to [Ne], as we already know. That is the reason, by the way, why chemical-bonding aficionados do not find $BaTiO_3$ terribly exciting, yet another dull ionic refractory – **unless** one highlights some underlying covalency as we will do now. Here, a look at Figure 4.8 will serve us well.

If it was only about point charges, a chemical-bonding picture of ionically interacting Ba^{2+}, Ti^{4+}, and O^{2-} would be satisfactory but already the strongly deviating charges of O (−0.96) and Ti (+1.13) prove that the idea is oversimplified,[8] possibly wrong; instead, there is a covalent Ti–O interaction with a bond order of 0.64, a quite substantial amount, also visible in the energy-resolved COBI plot. In stark contrast, Ba–O has almost no covalency (0.03), even lower than in the NaCl case seen before (Müller et al., 2021). Even though Ti has a higher oxidation state than Ba, the Ti charge density is larger, and taking into account Ti–O covalent interactions, $BaTiO_3$ is not so boring at all.[9]

That being said and rounding up things, it is easily possible, almost trivial to characterize whatever kind of solid-state material in terms of ionicity and covalency, at least without too many problems as long as they follow the octet rule, that is, all the individual

---

8 Here I have rounded the figures to two digits, which is quite sufficient. Everything is about **trends**.

9 Oxide perovskite systems **really** get interesting when the transition metal has leftover electrons (= has an open d shell) and becomes electronically active, such as in $BaFeO_3$ (Hoedl et al., 2022).

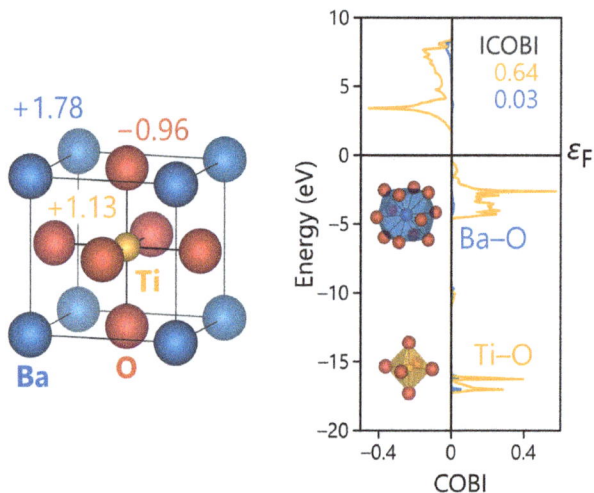

Figure 4.8: Unit cell of BaTiO$_3$ with Löwdin charges (left) and integrated/energy-resolved COBI (right).

atoms are striving for a noble-gas shell. Any kind of covalency indicator derived quantum-chemically and ionicity indicator also derived this way should serve us right, namely by "mapping" materials in such a derived covalency–ionicity space. By doing so, one may not necessarily increase chemical understanding, at least not finding unexpected things, but sometimes simply plotting materials helps in differentiating a wealth of chemical compounds. For example, one may use an averaged COHP integral (showing how much covalency contributes to the band-structure energy) as a covalency indicator and the Löwdin-derived ionicity (see above) for the latter, and this is what we plot in Figure 4.9.

The extremes of covalency (diamond and graphite) devoid of ionicity are found in the upper left corner, whereas the ionicity prototypes (fluorides of alkali and alkaline-earth metals) without significant covalency reside in the lower right corner; for a given cation, those phases involving higher homologues of the anions become less ionic. Metals are characterized by small covalency and small ionicity, and ionicity is zero for an element (such as copper), of course. If covalency is to increase, one arrives at weird semiconductors with small ionicity and moderate covalency (some of them violating the octet rule, these will be covered in Section 4.7) but regular semiconductors such as Si, Ge, and GaAs are even more strongly covalent (Nelson et al., 2023). Not to forget, there are certain examples with both significant covalent and ionic interactions, such as oxides in maximum oxidation states, both for transition metals (say, the volatile molecular oxide OsO$_4$) and for main-group elements (say, the already covered rock-solid SiO$_2$ in the quartz polymorph).

Because SiO$_2$ and all its polymorphs are particularly important for the earth sciences, we take it as a final example of this chapter. People have tried to group those many SiO$_2$ polymorphs in terms of the underlying ionicity, empirically and successfully so (Reddy et al., 2005), namely by deducing the latter properly from the optical refractivities depending on the ionic polarizabilities, thereby mirroring the ionic nature of the bond. And this we show in Figure 4.10, with a special focus on the quantum-chemically derived numbers plotted on the left.

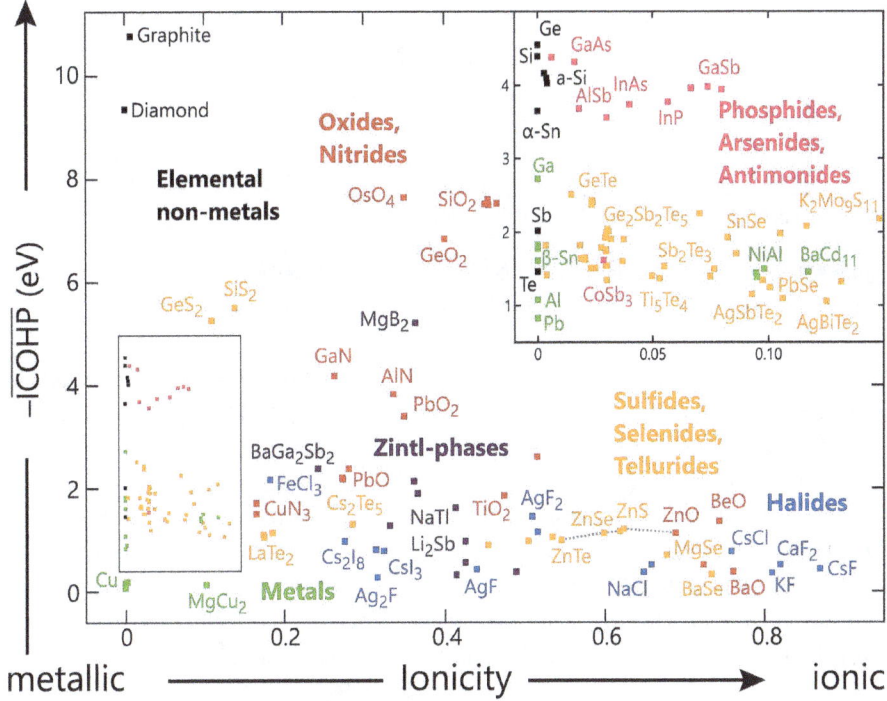

Figure 4.9: Covalency (expressed through an average ICOHP) and ionicity (derived from Löwdin charges) for a large number of solid-state materials based on DFT calculations.

These values are structure-dependent because – although depending on the same atoms, Si and O – they **vary** from polymorph to polymorph, in good accordance with the empirically derived values on the right. So, the polymorph's structure determines the Si–O bond lengths that affect the charge transfer and, hence, the ionicity. And vice versa, of course! The ionicity of α-quartz is lower than the one of β-quartz because the former has slightly longer Si–O bonds. For the next three polymorphs – α-cristobalite, β-cristobalite, and α-tridymite – the Si–O bonds are progressively becoming shorter and ionicity grows. The more dense SiO_2 polymorphs have wider Si–O bonds such that the ionicity must fall off, coesite even resembling α-quartz in terms of ionicity. Eventually, stishovite with the longest Si–O bonds is the least ionic (Esser et al., 2017).

The reader may be interested to learn that crystal chemistry has a long and proud (possibly too proud) history trying to interpret structural stabilities of complex ionic

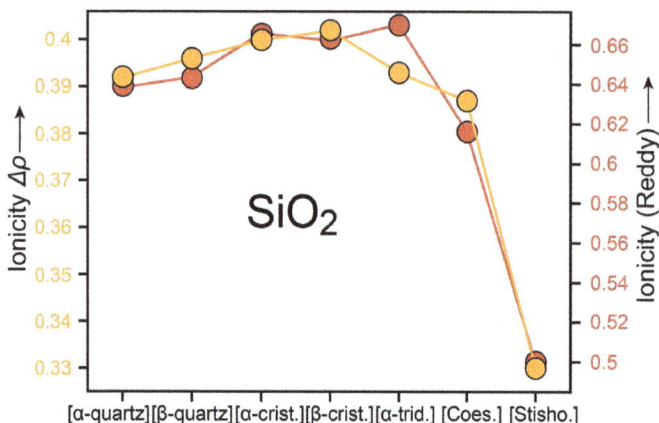

Figure 4.10: Comparison of quantum-chemically (left) and empirically derived (right) ionicities of seven SiO$_2$ polymorphs, the former based on prior DFT and charge calculations.

crystals (including silicon dioxide, of course) in terms of the famous Pauling rules, in particular, rule number 3. Here, the presence of shared edges, and especially of shared faces, in a coordinated structure decreases its stability (Pauling, 1960; Dronskowski, 2005). And there are countless examples, even including structural varieties on crystal **surfaces** (Deringer & Dronskowski, 2014), not only bulk materials, speaking in favor of the validity of the Pauling rules, at least at first sight, in particular for the third rule that does carry meaning. Admittedly, a statistical assessment of the performances of all Pauling rules **together** carried out over several thousands of oxides, however, does show the limits of those rules, also manifesting that only 13% of those oxides simultaneously satisfy the last four rules, thereby relativizing their predictive power (George et al., 2020). Nonetheless, one must admire the early crystal chemists for having derived those rules using a very sparse database.

To recap this chapter, let us summarize what seems important in navigating through the structure and bonding of such presumably simple systems:

- To understand chemical bonding in (simple) solids, one is well advised to know the periodic table and the valence-electron count (VEC) of the individual atoms. If it is important for simple systems, it may be even more important for more complicated ones.
- One should always check whether or not the atoms involved strive for the octet rule and if this (using either the covalent limit, such as in C, or the ionic limit, as in NaCl) leads to noble-gas shells for all species. If so, interpreting chemical bonding may become rather straightforward. Reality is real, so expect something **between** the covalent and ionic limits.
- The amount of covalency is easily visible from energy-resolved COHP plots (yielding the bond's share of the band-structure energy). It may be even easier to quantify from COBI and, most importantly, ICOBI values that directly yield the bond order. This almost looks like a matter of taste.

- Likewise, wave function-derived atomic charges should complement the aforementioned tools because both Mulliken and Löwdin charges stem from the same quantum-mechanical basis. Charges of that kind are important, at least as long as they make chemical sense.
- Such measures of covalency and ionicity compare well with traditional ways to plot or describe materials as a function of their individual bondings. For polymorphs, this has been empirically done before, successfully so, but it is also accessible, in modern times, based on electronic-structure theory only.

4.2 Mixed anions, complex anions, and complex cations

Upon recalling the content of Section 4.1 and the structures and bondings of supposedly simple systems obeying the octet rule, compounds such as NaCl, MgO, Sc_2O_3, GeO_2, and GaN look almost trivial to us, even if we were to include species with transition metals in "uncritical" oxidation states such as TiO_2, MnO, and FeN, in which the d shells are emptied or half-filled. And if one can make NaCl and KCl, then one can surely make the so-called solid solution $Na_{1-x}K_xCl$ (yes, it is possible). Likewise, the existence of MgO and CaO strongly suggests the existence of $Mg_{1-x}Ca_xO$ (another easy case), and one should also try to make $Mn_{1-x}Fe_xO$ given the knowledge of MnO and FeO. Such rather standard, even slightly boring solid-state chemistry by **substituting** one metal by another metal (sometimes incorrectly dubbed **doping**[10]) is often done because it is quite simple and allows to somewhat modify the physical and chemical properties, simply because the cationic (i.e., electron-providing) metal atoms are involved, and the occupied levels in the valence region are almost untouched because they are still held by the anions, in the above examples, chloride or oxide or nitride anions; whether Na or K provides an electron into the occupied levels below the Fermi energy, this does not much affect the chloride levels, let us put it this way.

Chemically substituting the **anions**, however, is far (far!) more difficult because we are now manipulating the filled valence levels directly. Hence, there is an enormous synthetic challenge but it is worthwhile accepting because such anion-modified materials will immediately show physical properties probably never seen before, this being the reason why **modern** solid-state chemistry is involved in doing that. In particular, nitride (opposed to oxide or halide) solid-state chemistry has become popular, despite the aforementioned synthetic challenges, for very good reasons (Niewa & Jacobs, 1996; Niewa & DiSalvo, 1998; Höhn & Niewa, 2017; Meissner & Niewa, 2021). But there is no reason to abolish the oxide anion completely, one may proceed by using

---

10 If one atom gets replaced by another one, this is called **substitution**. People often use the term **doping** if the additional atom entering the scene comes in an extremely small quantity, and it depends on thermodynamics and kinetics if the additional atom replaces an existing atom or goes into an unoccupied interstitial site.

smaller steps. An excellent example starts with titanium dioxide, TiO_2 (= $Ti^{4+}(O^{2-})_2$), the most important white pigment with numerous applications. To make the material more covalent, it is reasonable to replace half of the oxygen atoms by nitrogen and, in order to still have an [Ar] noble-gas shell for the metal atom, move from the tetravalent Ti toward the pentavalent Ta. Hence, TaON (= $Ta^{5+}O^{2-}N^{3-}$) is our target, a so-called oxyni-tride,[11] one representative of a constantly growing material class (Tessier, 2017).

Figure 4.11 provides structural insight of the ground-state polymorph of TaON, the so-called β-phase of the yellow-colored (not white anymore, hence, more cova-lency, as qualitatively predicted) baddeleyite structure type which looks rather weird due to the sevenfold coordination of Ta by O and N – I told you, mixed-anion species are unusual. The ugly coordination motif and the connectivities turn into a really messy crystal structure, so we only show the central polyhedron. The low symmetry translates into a rather spaghetti-like band structure (derived from the local-density approximation (LDA) in this case) and also a corresponding DOS in which the valence levels are mostly anion-centered, as expected, with a too small LDA bandgap of less than 2 eV. The (direct) optical gap is about 2.4 eV. The COHP immediately indicates less covalency in Ta–O than in Ta–N interactions, simple to tell – just by looking – from the integral of the two curves. The plot shown is from an old-style all-electron LMTO (linearized muffin-tin orbital) calculation using numerical local orbitals (called partial waves) which nicely get the physics and chemistry right, at least semiquantita-tively (Dronskowski, 2005).

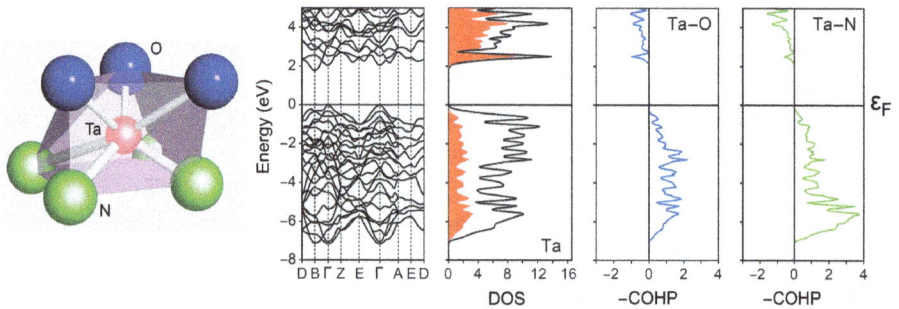

Figure 4.11: Crystal-structure motif of baddeleyite-type β-TaON (left) and electronic bands, DOS, and Ta–O and Ta–N COHP analysis (right) based on the LDA.

Given modern electronic-structure theory, it is rather straightforward to then move ahead and check for other phases, either reported or imagined. For example, a so-called α-phase can be immediately excluded by means of theory (Lumey & Dronskow-

---

11 This naming convention is controversial, in particular in German-speaking countries because they insist that "oxidonitride" or "oxide nitride" would be more appropriate. The somewhat ugly "oxyni-tride" is an established term, though, so we also use it here, with apologies.

ski, 2003), and one can also easily predict related high-pressure phases (Lumey & Dronskowski, 2005), for example, the cotunnite type of TaON with ninefold Ta coordination, prepared shortly after its prediction (Woodhead et al., 2014).

Similar calculations as regards the relative stabilities of TaON adopting the baddeleyite, anatase, rutile, and fluorite phases are nowadays routine, in particular when it comes to the anion (O, N) distribution over the available sites since this particular information is extremely difficult to achieve experimentally as long as (expensive) neutron diffraction has not been carried out (Bredow et al., 2006). In addition to that, theory can also help to identify new phases, for example, in fluorite-type TaON substituted with Y. While one often finds asymmetric reflection profiles in experimental X-ray diffraction patterns of supposedly cubic phases, this phenomenon simply mirrors the true, much lower symmetry, even triclinic, as evidenced by molecular-dynamics simulations of such matter. The materials may look cubic macroscopically but they are not like that microscopically, the reason being chemical bonding (please see below). In some cases, the tendency for anion ordering strongly depends on the temperature (Wolff et al., 2006).

A true synthetic breakthrough is given by the clever ammonolytic preparation of the metastable phase γ-TaON starting from Ta_2O_5 (Schilling et al., 2007), whose structure and COHP are given in Figure 4.12. This orange-colored material crystallizes in the monoclinic $VO_2(B)$ structure type, with octahedral coordination of the Ta atoms.

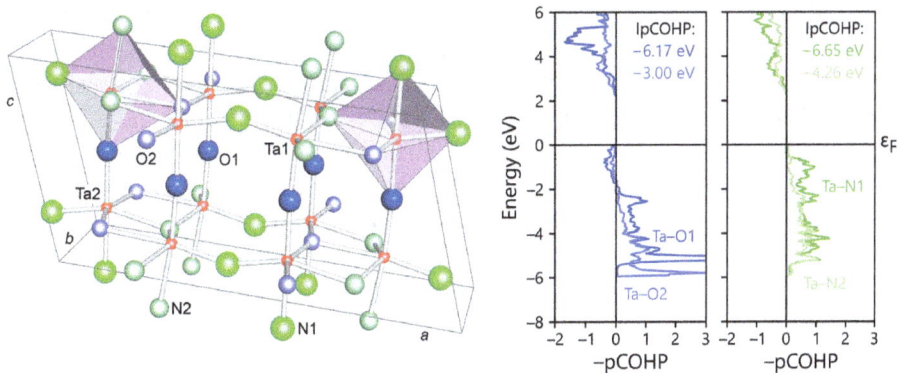

Figure 4.12: The crystal structure of γ-TaON showing the distribution of O and N atoms over four possible sites (left) and the associated COHP (right) projected from PAW plane-wave theory.

As alluded to already, neutron diffraction on small samples is either difficult or impossible to carry out,[12] so the atomic **assignment** (which anion is O and which is N?) and subsequent structural refinement by X-ray diffraction were only possible by quantum-

---

12 This does not question neutron diffraction, one of the finest (probably **the** finest) method to structurally and also magnetically characterize crystalline matter (Meven & Roth, 2017).

chemical means. The monoclinic material transforms irreversibly to β-TaON above 900 °C, and it turns out as being slightly less stable, close to 20 kJ mol^{-1} compared to the baddeleyite type. This lower stability is mirrored by the much lower density of the open γ-polymorph (8.6 g cm^{-3}) compared to the ground-state structure (11.0 g cm^{-3}) and its 35% smaller bulk modulus. Please note that such criteria may be misleading, though, because denser does not necessarily mean more stable. For example, high-pressure CaO is more dense but less stable than regular CaO (Appendix B).

As regards the chemical bonding, oxygen and nitrogen are occupying four different anion sites, marked as O1, O2, N1, and N2 in Figure 4.12. The corresponding coordination numbers are increasing from 2 for O1 to 3 for O2 and also N1 up to 4 for the N2 site (Wolff et al., 2007). As also shown in Figure 4.12, the Ta–N chemical bonding of both N1 and N2 is perfectly covalent up to the Fermi level, and the integrated projected COHP values indicate that it is stronger than Ta–O bonding, both O1 and O2, as also witnessed for the ground-state β-polymorph. This being a metastable phase, we also witness filled **antibonding** Ta–O levels just below the Fermi level, the reason for which we will clarify below.

And yet, a so-called δ-TaON polymorph can also be made, now by ammonolysis of amorphous Ta$_2$O$_5$ under slightly different conditions, and it leads the δ-phase to crystallize in the tetragonal anatase structure type, transforming irreversibly into the baddeleyite ground state (Lüdtke et al., 2014). Once again, Ta is octahedrally coordinated, and a quantum-chemical supercell calculation yields an anion distribution with maximum N–N distances. This immediately alludes to an electrostatic interpretation because the formally charged N^{3-} anions should repel each other but we nonetheless note that the covalency in the material is significant, as a careful look into the bonding corroborates; the electrostatic model is a useful model but still a model.

For illustration, Figure 4.13 provides an energy–volume plot of the three most important TaON phases, together with a finite-temperature Gibbs energy diagram calculated from the quasi-harmonic phonons. The plot in Figure 4.13 (left) confirms that β-TaON is the ground-state structure, and the nearest polymorph in energy is given by the γ-phase. The δ-phase is even higher in energy than γ but could be derived from the latter by applying some pressure due to δ-TaON's volume being slightly smaller. In terms of Gibbs energies (right),[13] β-TaON should transform into γ-TaON slightly above 500 K but δ-TaON can **never** be reached using thermodynamic control. A toast to the skillful experimentalists who succeeded in making δ-TaON, nonetheless, using kinetic means.

Coming back to optical properties and the proximity to TiO$_2$, it is obvious that materials such as TaON should be promising in photocatalysis, for example, light-induced

---

[13] As regards thermodynamics, there is the internal energy U which corresponds to the total energy E from electronic-structure theory. If we include volume work pV, we get to the enthalpy $H = U + pV$. By also including temperature T and entropy S, we eventually arrive at the free enthalpy or Gibbs energy $G = H - TS$.

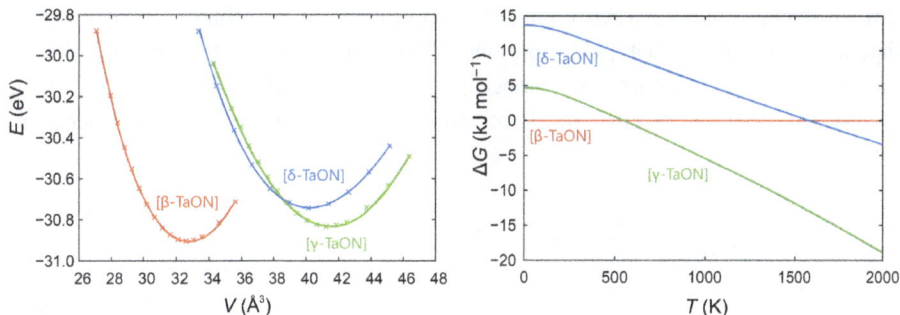

Figure 4.13: Density-functional energy–volume diagram of β-, γ-, and δ-TaON (left), and Gibbs energy diagram as a function of temperature (right).

water splitting. Given sufficiently precise quantum-chemical methods, that is, GW quasi-particle energy calculations (in combination with the plasmon-pole GW approximation) one can even screen the direct (indirect) optical bandgaps that arrive at 2.99 (3.20) eV for β-TaON, 2.38 (2.59) eV for γ-TaON, 2.52 (3.01) eV for δ-TaON, 1.85 (2.21) eV for rutile-type TaON, 1.78 (2.50) eV for fluorite-type TaON, and eventually 1.57 (1.77) eV for tetragonal-ZrO_2-type TaON (Lüdtke et al., 2017). As such, the bandgaps decrease with decreasing polymorphic stability, yet another corroboration of the principle of maximum hardness (Pearson, 1997).

We already alluded to the chemical-bonding competition between oxygen and nitrogen atoms found in oxynitrides, the oxygen/nitrogen atoms striving for more/less ionicity and less/more covalency, a consequence of the higher/lower EN of oxygen/nitrogen. This can be easily demonstrated by an oxynitride variant of the $BaTiO_3$ perovskite known from Section 4.1, and our target phase is now called $BaTaO_2N$ (= $Ba^{2+}Ta^{5+}(O^{2-})_2N^{3-}$), similar to the process of changing TiO_2 into TaON. Clearly, $BaTaO_2N$ is expected to also crystallize in a structure resembling perovskite, and this is corroborated both from experiment and electronic-structure theory but only to a certain degree. Theory immediately shows that a perfectly cubic crystal structure looks slightly unfitting for $BaTaO_2N$, as depicted in Figure 4.14 (left). Just below the Fermi level, there are unfavorable antibonding Ta–O levels for cubic $Pm\bar{3}m$ symmetry, and they are removed upon lowering the symmetry to the orthorhombic system, space group $Pnma$, because **this** allows for different positions for O and N coordinating the central Ta atom. Supercell calculations even suggest space group $Pmc2_1$ for optimized bonding. The effect is so fundamental that it can even be illustrated from a molecular model, namely, $[Ta(OH)_4(NH_2)_2]^-$ of which some semiempirically calculated molecular orbitals just below the HOMO are depicted in Figure 4.14 (right). By enforcing the same Ta–O and Ta–N bond lengths, there **must** be antibonding Ta–O interactions (in the MO plots, at −0.47 and −1.01 eV), and this is what drives anion ordering in these oxynitrides. O and N simply need different positions to optimize their ionicity/covalency proclivities, thereby making the Ta–O bonds longer (more ionicity) and the Ta–N bond shorter (more covalency) (Wolff & Dronskowski, 2008). We reiterate that covalency is an interference phenomenon of overlapping orbitals, so the distance should be right; for ionicity, it simply falls off with the inverse distance. And all that leads to a very rich chemistry and associated material properties (Lerch et al., 2009). In γ-TaON discussed before, such Ta–O antibonding levels persist even in the experimentally reported structure.

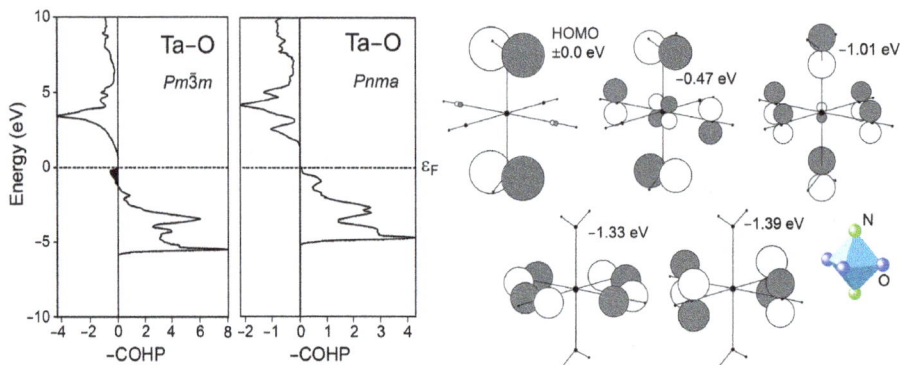

Figure 4.14: Tantalum–oxygen COHP bonding analysis in two structure types of $BaTaO_2N$ (left) as well as molecular-orbital plots in the proximity of the HOMO of a singly charged $[Ta(OH)_4(NH_2)_2]^-$ model system (right). In the MO representation, the MO energies are relative to the HOMO at 0.0 eV.

There are plenty of other mixed-anion combinations, and for reasons of brevity we simply mention three fundamental ones, nitride halides, oxyhalides, and also oxide hydrides (Zhang et al., 2017), the latter not to be confused with hydroxides. For example, the phase dubbed Mg_2NCl is accessible from a solid–solid reaction between Mg_3N_2 and $MgCl_2$, and the structural motif of the hexagonal crystal structure is depicted in Figure 4.15 (left), with layers of edge-sharing NMg_6 octahedra and Cl^- anions occupying the voids between them; the central Mg cation is octahedrally coordinated by both N and Cl anions. As regards bonding seen in Figure 4.15 (right), the more covalent Mg–N interactions prevail, and only the less covalent (by about $1/6$) but more ionic Mg–Cl interactions have to tolerate a little antibonding below the Fermi level (Li et al., 2015). Whether or not such material can also be used for optical applications is unclear as yet, but the rather expensive HSE06 functional yields a direct bandgap of about 3.6 eV. And there is more chemistry of that kind.[14]

Likewise, a similar fine adjustment of covalency is achieved in the much more complex oxyhalide phase Bi_2YO_4Cl of which there are also variants in which Y is re-

---

[14] Similar, comparatively simple compounds exist, Zn_2NX (with X = Cl, Br, I) being typical examples (Liu et al., 2013) which allow to chemically fine-tune the bandgap. The very courageous chemist may even prepare explosives such as PbN_3X (with X = Cl, Br) in which the halide anion is accompanied by the **azide** anion, N_3^-, a rather metastable species with a tendency for decomposition (Liu et al., 2020).

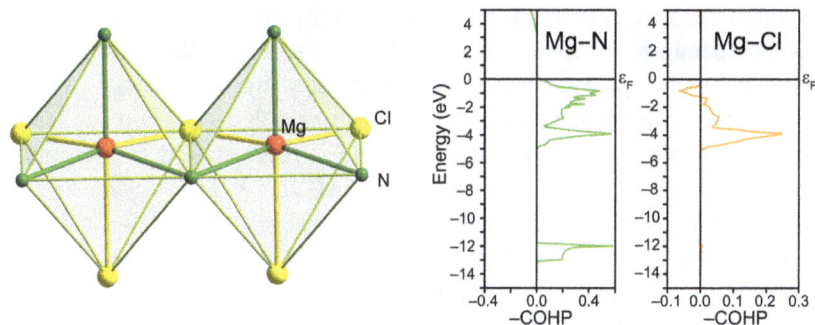

Figure 4.15: Structural motif from the crystal structure of Mg_2NCl (left) and corresponding Mg–N and Mg–Cl COHP bonding analysis (right).

placed by La or Bi (Nakada et al., 2021). Such materials containing triple-fluorite Bi_2YO_4 blocks separated by chlorine anions are made using flux methods, and they show excellent photoconductivity due to both a narrow bandgap and, at the same time, a highly dispersive structure of the conduction bands. This leads to visible-light catalytic activity for water splitting, thanks to the crystal and electronic structure given in Figure 4.16, in particular for the Bi_2YO_4Cl phase which is special.

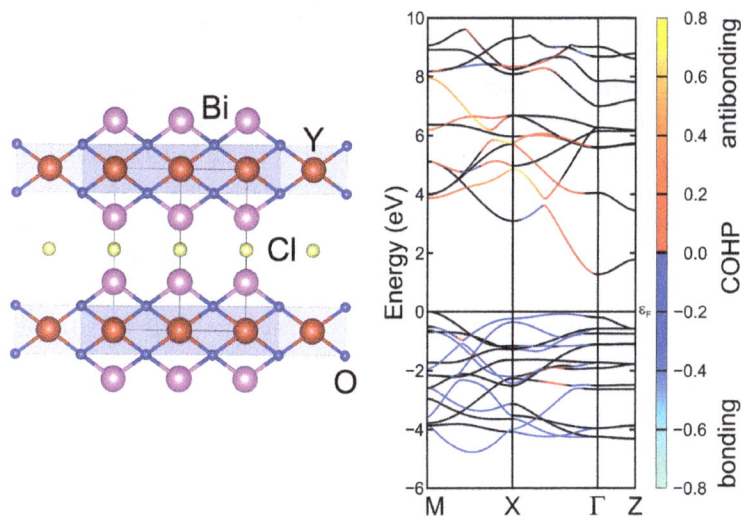

Figure 4.16: The crystal structure of Bi_2YO_4Cl (left) and the corresponding band structure (right) color-coded according to the COHP bonding interaction between Bi $6p_z$ and O $2p_x$ orbitals, nonbonding at Γ.

As shown in Figure 4.16 (right), the course of the conduction band is highly dispersive, thanks to the changing Bi–O bonding interaction as a function of the course through reciprocal space. Note that the required dimensional control of the crystal structure can only be accomplished by using at least two anions at the same time. Bi_2YO_4Cl has an experimental bandgap of 2.50 eV, based on UV–vis diffuse reflectance data, thanks to a downshifting conduction band; hence, it shows higher photoconductivity than most related photocatalysts for water splitting. The related phase $Bi_{12}O_{17}Cl_2$, also an oxyhalide, contains an even sextuple Bi–O layer composed of rock salt and fluorite units. Here, the valence-band maximum is controlled by the O $2p$ orbitals, as evidenced by local densities of states, so the stability of this oxychloride against photogenerated holes is easily understandable (Kato et al., 2022). In any case, such spatially resolved bonding analysis in reciprocal space may be used by the clever experimentalist to rationally proceed synthetically, given the fact that we know the phases of the interacting orbitals. A plain density plot would not show the needed information.[15]

How about "complex" anions, and what do chemists mean by that? Complex anions contain two or more bonded atoms which **together** form the no-longer-simple (hence, complex) anion. One of the simplest examples would be given by barium peroxide, BaO_2, easily made in the laboratory. Because Ba is an alkaline-earth metal, its cation must be formulated as Ba^{2+}, and a higher oxidation state (Appendix F) is impossible under standard conditions. So, the complex peroxide anion reads O_2^{2-} or ^-O–O^-, with a single bond, just like in the isoelectronic Cl_2 molecule. BaO_2 adopts the CaC_2 structure type and poses no challenge for DFT, even its *ab initio* force constants can be accurately calculated, eventually yielding quantitative agreement between theoretical and experimental Raman spectroscopic data (Stoffel & Dronskowski, 2012). We will meet the nitrogen equivalents, so-called pernitrides and diazenides, in Section 4.6.

There are other complex anions needed to mimic and/or modify the chemistry of simple anions (Schädler et al., 1993). For example, one would chemically love to have a divalent nitride anion, so "N^{2-}" instead of N^{3-} as required by the octet rule and the [Ne] configuration, for example to somehow repeat, on purpose, oxide chemistry by replacing O^{2-} by the more covalent nitrogen. Because "N^{2-}" is unstable, however, the complex anion NCN^{2-} (which is the doubly deprotonated H_2NCN cyanamide molecule) serves as a convenient replacement, this complex anion **does** exist, with the same electron count as CO_2. For completeness we note that NCN^{2-} either comes with the so-called symmetric **carbodiimide** shape (when the central C engages in two double bonds to the left and right nitrogen, $[N=C=N]^{2-}$) or with the less symmetrical **cyanamide** shape (a single C–N bond to the left, a triple C≡N bond to the right, $[N–C≡N]^{2-}$). And then a huge number of binary (ternary and quaternary) phases can be made in which the covalent NCN^{2-} has replaced the more ionic O^{2-}. These novel compounds are not only more covalent than the corresponding oxides, as corroborated by experiment (Boyko et al., 2013) and theory (Nelson et al., 2017), they also allow for more structural degrees of freedom since the one-dimensional NCN^{2-}, in contrast to the zero-dimensional point-like O^{2-}, may

---

15 A similar strategy focusing on quantifying chemical bonding in reciprocal space may turn out highly useful in discovering thermoelectric chalcogenides by high-throughput material screening (Xi et al., 2018).

tilt and **twist**, hence form differently sized crystallographic sites to accommodate whatever kind of metal cation (Corkett & Dronskowski, 2019). Three characteristic binary phases involving transition-metal cations are displayed in Figure 4.17. From a crystallographic point of view, their so-called Bärnighausen (group–subgroup) trees are available (Pöttgen et al., 2023), showing how the different structures relate to each other.

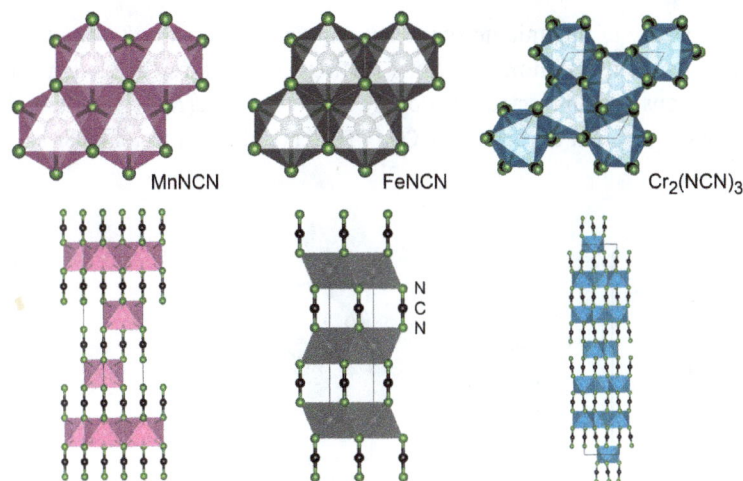

Figure 4.17: Structural motifs in MnNCN (left), FeNCN (middle), and $Cr_2(NCN)_3$ (right).

When transition metals are involved (see Figure 4.17), prominent binary examples are MnNCN (Liu et al., 2005), FeNCN (Liu et al., 2009), and $Cr_2(NCN)_3$ (Tang et al., 2010). The strongly correlated phase CuNCN (Tchougréeff et al., 2012) is a big problem for DFT (in terms of magnetism), and the two forms (carbodiimide and cyanamide) of HgNCN (Liu et al., 2003) are also real DFT challenges (Dronskowski, 2005). The simpler main-group binaries CaNCN (Vannerberg, 1962) and PbNCN (a cyanamide) (Liu et al., 2000) pose fewer problems. We will soon come back to these species.

Other complex nitrogen-based anions can also be deduced, so to speak, from the molecule **guanidine**, CN_3H_5, the all-nitrogen cousin of urea, CON_2H_4, and the mysterious carbonic acid, CO_3H_2 (see Section 4.6). Guanidine is difficult to crystallize (Yamada et al., 2009), but its deprotonated variants (shown in Figure 4.18 on the right) serve as nice synthetic entities. Under regular conditions, the very basic molecule guanidine (similar to KOH) readily accepts a proton and then turns into the **guanidinium** cationic species (to the left). Clever chemical synthesis, however, lets guanidine lose one (or more) proton(s), and then **guanidinate** anionic entities result (to the right).

That being said, complex salts between guanidinate complex anions and simple metals such as $RbCN_3H_4$ (Hoepfner & Dronskowski, 2011; Hoepfner et al., 2013) or $NaCN_3H_4$ or KCN_3H_4 (Sawinski & Dronskowski, 2012) or even $Ba(CN_3H_4)_2$ (Benz et al., 2019) can be made, and there are further examples involving 4f metals (Görne et al.,

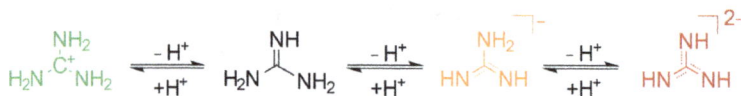

Figure 4.18: Protonated (guanidinium, left) and deprotonated (guanidinates, right) complex cations and anions as derived from molecular guanidine.

2016). The rich complexity of the anionic shape adopting *syn, anti,* and *all-trans* conformations depending on the H-atom positions further complicates the structural chemistry. If two protons are subtracted, the doubly deprotonated $C(NH)_3^{2-}$ dianion is formed, the nitrogen equivalent of carbonate, CO_3^{2-} The crystal structure of its Sr salt is shown in Figure 4.19.

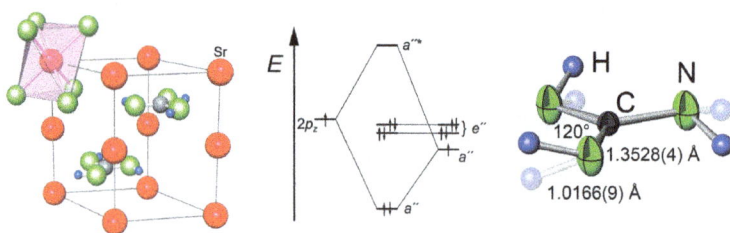

Figure 4.19: The crystal structure of $SrC(NH)_3$ (left) and the anionic shape (trinacria motif) of the doubly deprotonated guanidinate anion (right) based on neutron data, together with its molecular-orbital diagram for the central C $2p_z$ interacting with N $2p_z$ ligand-group orbitals (middle).

While Sr is octahedrally coordinated by nitrogen atoms, the CN_3 core of the trinacria-shaped complex anion is fully planar with C_{3h} symmetry, and the H atoms are slightly above or below the mirror plane. The strange C–N bond length of 1.35 Å mirrors a bond order of 1⅓, a consequence of the aromatic character of the dianion or, alternatively, expressed, of a π bond delocalized over the C–N triangle (Missong et al., 2015). And similar chemistries are also developing from derivatives of other, related molecules, for example **melamine**, $C_3N_3(NH_2)_3$, the trimer of molecular cyanamide, H_2NCN. Melamine is the main precursor for the synthesis of graphitic carbon nitride (with an idealized formula of C_3N_4). The acid–base chemistry of melamine is sketched in the left part of Figure 4.20.

Accepting a proton to yield the melaminium cation is a rather typical (even boring) reaction of the base melamine. Within liquid ammonia, however, melamine may react as an acid, split off a proton, and form the **melaminate** anion which then binds to an alkali cation and, thus, defines a new compound class, solid-state melaminates (Görne et al., 2021). By means of first-principles theory one can also predict existence, synthetic conditions, and whatever properties of a fully hydrogen-free melaminate **salt** dubbed WC_3N_6 with W^{6+} besides $C_3N_6^{6-}$ depicted in Figure 4.20 (right). The layer-like porous material is stabilized by proper W–N covalent bonding in addition to a huge Madelung field. Likewise, the HSE-predicted bandgap of 2.25 eV and suitable band-edge potentials

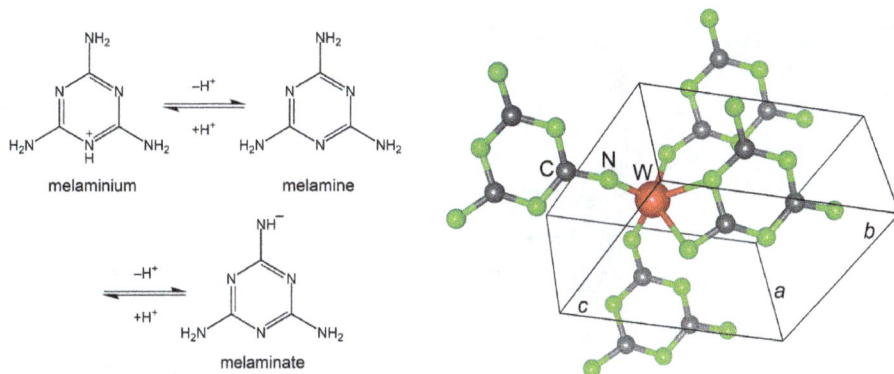

Figure 4.20: Neutral melamine (left) and its acid–base chemistry to yield a protonated melaminium complex cation or a deprotonated melaminate complex anion, together with a totally deprotonated melaminate salt of tungsten, WC_3N_6 (right).

suggests its use in photocatalysis (Chen et al., 2023), probably superior to graphitic C_3N_4. The complex anion $C_3N_6^{6-}$ makes the difference.

But let us return to the aforementioned NCN^{2-} carbodiimides and mixed-anion species based on such "divalent nitrides." Besides the already discussed oxynitrides, nitride halides, and oxyhalides, there should also be carbodiimide halides, why not? And there are, for example, compounds such as $Hg_3(NCN)_2Cl_2$ crystallizing in the acentric space group $Pca2_1$ with fourfold and fivefold Hg coordination by N and Cl. The direct bandgap is about 3.1 eV, and the electronic-structure theory highlights the high potential of second-harmonic generation (SHG) due to large SHG coefficient and birefringence values, both caused by the carbodiimide group (Qiao et al., 2021). Another example is the red oxide carbodiimide $Sn_2O(NCN)$ which looks like an atomically mixed intergrowth between SnNCN and SnO, layer by layer, and the tin is divalent throughout, therefore **deviating** from the octet rule; we will come back to such phenomena in Section 4.7. For the moment being, a look at Figure 4.21 helps us understand the structural chemistry of the material (Dolabdjian et al., 2018) and its bonding.

In the orthorhombic crystal structure, the local coordination of Sn^{2+} mirrors the presence of two additional electrons, often dubbed as the electron "lone pair" on Sn^{2+} not residing in tin's maximum oxidation state (+IV), with fourfold coordination by N and O. Theoretically, the phase is an indirect semiconductor with a bandgap of about 1.6 eV by use of the MBJ exchange-correlation functional. The Sn 5s orbital is strongly involved in covalent bonding (see Figure 4.21), also detectable from a fatband analysis (not shown). The Sn–O and Sn–N bonding can be nicely differentiated, and there are occupied antibonding levels, thanks to the $5s^2$ configuration. As expected, C–N bonding in the carbodiimide unit is strong and perfectly optimized, also seen from the ICOHP values. So, the valence band mostly consists of Sn 5s, O 2p, and N 2p levels, while the conduction band is dominated by Sn 5p levels.

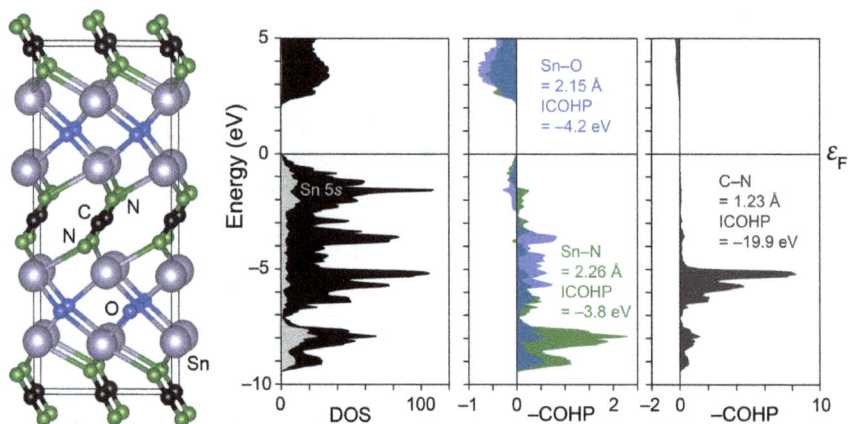

Figure 4.21: Crystal structure of $Sn_2O(NCN)$ on the left as well as DOS and chemical-bonding COHP analysis in $Sn_2O(NCN)$ on the right.

$Sn_2O(NCN)$'s exciting electronic structure reveals the phase, through Mott–Schottky experiments, as an n-type semiconductor with a flat-band potential of −0.03 V such that the valence-band edge alludes to photochemical water oxidation. In fact, $Sn_2O(NCN)$ almost doubles the photocurrent of $CuWO_4$ thin films (Chen et al., 2020). A related bandgap tuning is also seen for the oxide carbodiimide Bi_2O_2NCN crystallizing in the tetragonal antitype of $ThCr_2Si_2$ (taking the NCN^{2-} anion as a building block), a well-known structure type. Bi_2O_2NCN involving Bi^{3+} (and a low-lying $6s^2$ configuration due to relativity, see Appendix G) also deviates from the octet rule (see Section 4.7) and is an indirect semiconductor with a bandgap of 1.4 eV (theory) and 1.8 eV (experiment) (Corkett et al., 2019). Its crystal structure, displayed in Figure 4.22 (left), shows eightfold coordination of Bi by O and N (square antiprism), and also eightfold coordination (cube-like) of the carbodiimide group by Bi.

The indirect character of this semiconductor is easy to recognize in the band structure (between the special points Γ and A); see Figure 4.22 (right). A closer analysis shows the conduction-band minimum to be Bi $6p$ in character, and the N $2p$ levels dominate the valence-band maximum. Because of the slightly larger EN of N compared to S, the bandgap of Bi_2O_2NCN is larger than in Bi_2O_2S, another existing mixed-anion phase, this time an oxysulfide.

Instead of continuing with yet another complex anion being composed of two or more simple anions (an endless story), one might well ask the question if complex **cations** exist as well. In fact, there is one prominent and very recent example, namely from nitride chemistry, and it deals with an intellectually highly pleasing phase, namely, $Pb_2Si_5N_8$. It is the first nitridosilicate containing the highly electron-affine Pb^{2+} cation (Bielec et al., 2019), and it must be synthesized by a clever ion-exchange metathesis between lead chloride, $PbCl_2$, and a related strontium silicon nitride, $Sr_2Si_5N_8$. That is to say that Pb is too

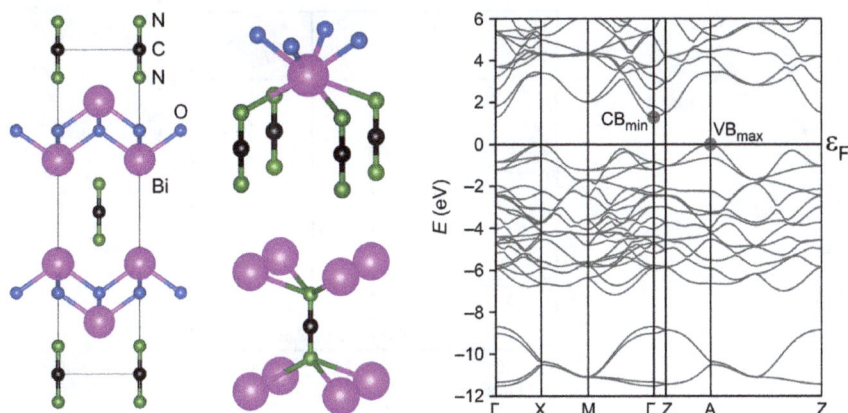

Figure 4.22: Crystal structure of Bi_2O_2NCN (left), Bi and NCN coordination (center), and electronic band structure (right).

electronegative to be directly oxidized by nitrogen, and this is the reason why there is no **stable** binary phase between Pb and N but explosive lead azide, $Pb(N_3)_2$.

After synthesis, structural and vibrational characterization, three puzzling details came to the surface: first, the Pb–N bonds show a rather broad spectrum, far more diverse than in other cases (with Pb being replaced by Sr or Ba, for example). Second, there is a likewise short Pb^{2+}–Pb^{2+} distance of 3.19 Å which might indicate some attractive interaction, however, going against all expectations. Third, a novel Raman signal at 117 cm^{-1} is observable, after some time identified (by phonon calculations) as going back to the vibration of a Pb_2 dumbbell charged +4, not at all expected. How could such a complex cation being composed of two Pb^{2+} ions stick together? Spin–orbit coupling as a consequence of relativity (see Appendix G) does not play a role. Figure 4.23 shows the crystal structure (left) and also the DOS and Si–N and Pb–N bonding analysis by COHP.

There is an obviously short Pb–Pb contact, and the metal–nonmetal bondings (Si–N and Pb–N) are strong, so the system is highly covalent to begin with. The electronic situation in a Pb–Pb dumbbell charged +4 is sketched in Figure 4.24.

Formally, the $6s^2$ configuration of Pb^{2+} corresponds to the electronic situation of [He], and a He_2 dimer is plainly impossible because in the iconic MO diagram of the hydrogen molecule depicted in Figure 4.24 (left), both bonding $1\sigma_g$ and antibonding $1\sigma_u^*$ would be filled, so He_2 would explode into two He atoms, and the same should happen for a [Pb–Pb]$^{4+}$ dumbbell. Given a highly covalent matrix, however, further sp mixing sets in,[16] and the bonding pattern changes from bonding-antibonding-bonding-antibonding (no sp

---

16 In an isolated Pb atom or Pb^{2+} ion, the $6s$ and $6p$ atomic orbitals are orthogonal to each other. The [Pb–Pb]$^{4+}$ unit is quasi-molecular, however, so s ($l = 0$) and p ($l = 1$) are no longer "good" quantum numbers, and this makes make these orbitals mix.

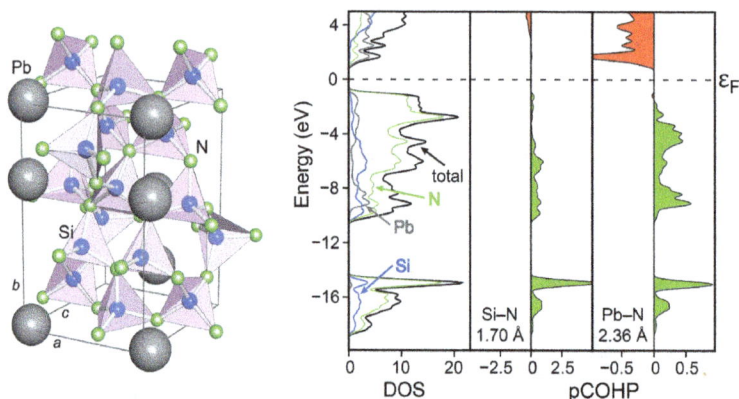

Figure 4.23: Crystal structure of $Pb_2Si_5N_8$ (left) as well as DOS, and Si–N and Pb–N chemical-bonding analyses (right), as projected from PAW plane-wave theory.

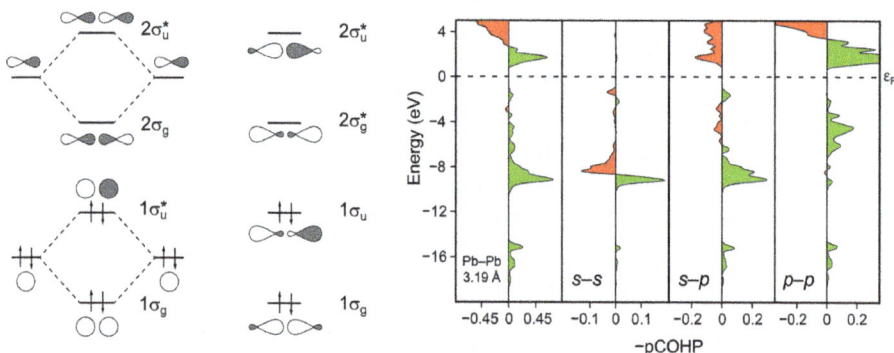

Figure 4.24: Schematic molecular-orbital diagram of σ-type interactions between two adjacent Pb^{2+} without (far left) and with *sp* mixing (left) and COHP analysis of Pb^{2+}–Pb^{2+} bonding involving different orbital types on the right.

mixing) to strongly bonding-weakly bonding-weakly antibonding-strongly antibonding (strong *sp* mixing) such that the two bonding lower levels host the four electrons of the $[Pb–Pb]^{4+}$ dumbbell, resulting in a complex **cation**. This back-of-the-envelope argument (Hu et al., 2012) is fully supported by the orbital-resolved DFT bonding analysis in Figure 4.24 (right). Simple *s–s* interactions alone would give no effect but significant *p* mixing into *s* makes all the difference (Bielec et al., 2019). I guess we will see more like that in the future.

To recap this rather lengthy and pretty "chemical" chapter, what needs to be remembered whenever mixed anions and complex anions (and cations) are to be dealt with?

- Mixed anions increase the chemical complexity in solid-state systems, and there is an immediate competition among the anionic species striving for an optimum of covalency and ionicity, thereby determining observables such as bandgaps.
- Electronic-structure theory and subsequent bonding analysis are powerful tools to get the **local** structures right, in particular for anions with similar scattering power which are difficult for X-ray structure determination.
- Complex anions chemically derive from molecular precursors, for example, by simple deprotonation; hence, carbodiimides, guanidinates, melaminates, etc. come to existence (with an emphasis of N-containing anions, attractive due to higher covalency).
- Solid-state materials with optocatalytic properties may be fine-tuned by proper chemical engineering, in particular when the chemical bonding is analyzed in reciprocal space by COHP-colored bands, and there are plenty of possibilities.
- For very covalent systems, there is even a chance for complex cations given proper orbital mixing.

4.3 Covalency versus ionicity in battery materials

In the two preceding chapters, we have witnessed the importance of both covalency, for example, in the case of a semiconductor (diamond/silicon), and of various amounts of ionicity, in particular for inorganic solids. A growing amount of ionicity simply makes materials more salt-like, and then covalency and ionicity can be fine-tuned, so to speak, by mixed anions, by complex anions, and so forth. This rather brief chapter, however, shall deal with a class of materials in which the interplay of covalency and ionicity not only determines the structures of the materials but also affects their **reactivities**. Because of the importance of such materials usually described as being electrochemically active, battery materials are truly interesting to study, and the enormous, almost incredible hype with everything that relates to "batteries" will not change that. Please note that this chapter is **not** meant as a mini-review for battery materials because there are **much** better ones (McCalla, 2017; Li et al., 2018b), accompanied by a sheer plethora of research papers covering the subject. In what follows, I just want to highlight a few bonding aspects of this admittedly fascinating area, nothing more.

We have already seen that Löwdin charges as derived from the wave functions are highly useful to quantify ionicity, say, in the simple halides of the alkaline metals such as NaCl and CsCl, depicted in Figure 4.25. To allow for a little diversity, let us look at slightly larger Mulliken charges, also derived like that.

The different plus–minus sizes of the Mulliken charges reflect the differences in ENs, CsCl being the more ionic phase than NaCl. A slightly less ionic (more covalent) phase would be given by LiCl but changing chlorine to nitrogen makes a real difference, in addition to a change in stoichiometry (Ertural et al., 2019). That being said, lithium nitride, Li_3N, the only stable alkali nitride known, is a fascinating material that "lives" from the

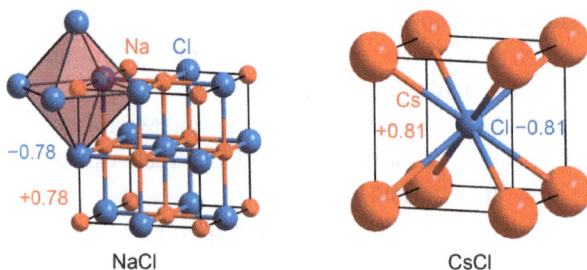

Figure 4.25: The crystal structures of NaCl (left) and CsCl (right), together with their wave function-based Mulliken charges, based on plane-wave DFT.

higher amount of covalency. Figure 4.26 shows the crystal structure of Li_3N, together with an experimental difference electron-density plot as a function of temperature.

The crystal structure is rather unusual for a supposedly ionic material, so this already indicates covalency; while the N atom is eightfold coordinated by Li, two Li are threefold coordinated by N, and one Li is twofold coordinated by N. Li_3N is an extremely good ionic conductor in the solid state, also called a **solid electrolyte**, and this behavior is easily illustrated by looking at the difference electron densities (Figure 4.26, right), indicating the "leftover" electron densities not accounted for by the X-ray refinement, derived upon subtracting the atomic electron densities from the experimental density. On purpose, pioneering X-ray data which were able to detect that weird phenomenon (Zucker & Schulz, 1982) four decades ago already are depicted. So, if there are residual densities (and there are, in particular when the material gets heated), the atoms cannot be resting still at their high-symmetry positions.

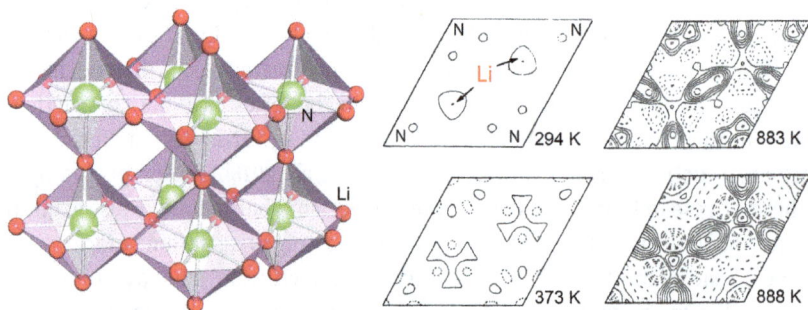

Figure 4.26: The crystal structure of Li_3N (left) and experimental difference electron-density plots as a function of temperature in the $z = 0$ layer (right).

Far below Li_3N's melting temperature of 1,086 K (or 813 °C), the Li substructure becomes quite mobile or even "melts" two-dimensionally, so the ionic mobility is extraordinarily high, an important ingredient for an electrochemically active material.

The electron density associated with Li somehow "smears out" as the temperature rises, so Li seemingly distributes in space, a typical behavior of an electrolyte. From a more theoretical perspective, the partial "melting" of the Li substructure is reflected by the Mulliken charges which are pretty small to begin with, only +0.23 for Li and −0.69 for N. Clearly, covalency is more important here. In more classical terms, the atomic movement from one site into yet another, previously vacant one corresponds to **defects** in the crystal structure, a highly important term needed to understand ionic conduction (Maier, 2017).

Before moving forward and exploring other related materials, let us remind ourselves that materials containing Li (and, to a lesser extent, Na) ions are at the core of rechargeable devices for energy storage. Somewhat simplified, the energy may be stored electrochemically via a reversible **intercalation** reaction such as

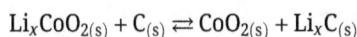

$$Li_xCoO_{2(s)} + C_{(s)} \rightleftharpoons CoO_{2(s)} + Li_xC_{(s)}$$

in which the Li ions move between two intercalating layered materials, a cathodic one such as $Li_xCoO_{2(s)}$ and an anodic one such as graphitic $C_{(s)}$. Graphite, the cheap layered allotrope of carbon, is decisive, and expensive diamond would be entirely useless; only the layered "two-dimensional" structure does the job. Admittedly, the electrochemical steps are far from being simple and also involve another in-between electrolyte, for example, $LiClO_4$ solvated in propylene carbonate. Unfortunately, Na does **not** intercalate well into graphite, so this reaction only runs with Li. On the other side, there are also **conversion** reactions, one example being

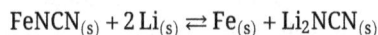

$$FeNCN_{(s)} + 2\,Li_{(s)} \rightleftharpoons Fe_{(s)} + Li_2NCN_{(s)}$$

in which the two chemical compounds are converted; solid-state chemists would call this a metathesis. Exactly the same reaction also runs with Na instead of Li without any problems. Here, the complex carbodiimide anion plays an important role by increasing covalency; it far outperforms the corresponding oxide, FeO. So, the synthetic challenge to work with less trivial anions pays off. At the beginning, the materials are crystalline, but as soon as the first discharging has happened, only amorphous phases are involved, even after recharging (Eguía-Barrio et al., 2016; Sougrati et al., 2016).

The presently far more abundant intercalation batteries include cathode materials that typically consist of metal oxides (such as the aforementioned $LiCoO_2$), silicates, or phosphates, while graphite-based compounds are used as anode materials. And now let us have a look at a few representative cathode materials and their charge distribution, calculated using DFT. Because transition metals (Co and Fe) are involved, the amount of electronic correlation (not correctly covered using a standard functional such as PBE) must be explicitly considered, so Co gets an extra +4.91 eV as a

Hubbard-like U correction,[17] whereas +3.71 eV is used for Fe. At the same time, these materials are known to exist in an antiferromagnetic (AFM) ground state, so AFM spin-polarized supercells are employed, and they also give proper bandgaps. Figure 4.27 allows for an overview of a few prominent inorganic solid-state compounds operating as cathode materials.

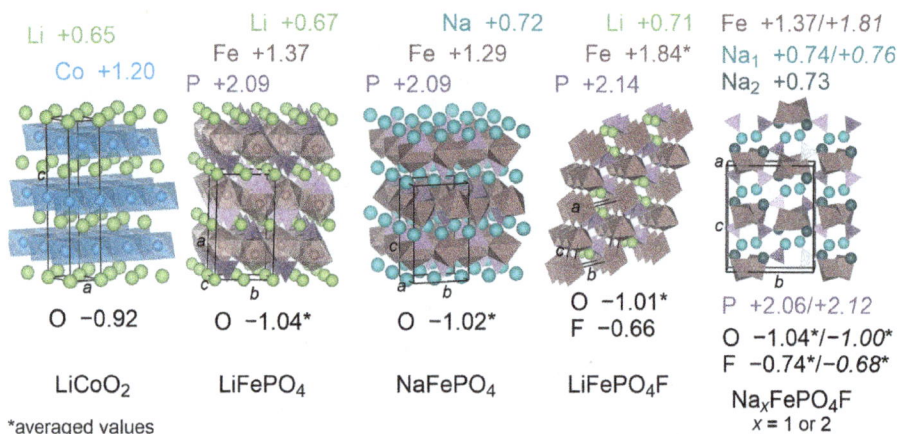

Li +0.65
Co +1.20
O −0.92
$LiCoO_2$

Li +0.67
Fe +1.37
P +2.09
O −1.04*
$LiFePO_4$

Na +0.72
Fe +1.29
P +2.09
O −1.02*
$NaFePO_4$

Li +0.71
Fe +1.84*
P +2.14
O −1.01*
F −0.66
$LiFePO_4F$

Fe +1.37/+1.81
Na_1 +0.74/+0.76
Na_2 +0.73
P +2.06/+2.12
O −1.04*/−1.00*
F −0.74*/−0.68*
Na_xFePO_4F
$x = 1$ or 2

*averaged values

Figure 4.27: The crystal structures of several representative cathode materials with wave function-derived Löwdin charges. In the case of Na_xFePO_4F, the charges relate to an Na content of 1 or 2.

In the present case, we take Löwdin charges to allow for an easy comparison with later materials in which they really excel against density-based charges. There is the layered $LiCoO_2$ with trivalent Co^{3+} and the ternary phosphates $LiFePO_4$ and $NaFePO_4$ (olivine type) with divalent Fe^{2+}, and the Löwdin charges do not vary much from different magnetizations or U parameters. Despite a smaller oxidation state, Fe is slightly **higher** charged than Co, and Na is higher charged than Li, a simple consequence of EN differences. It is often assumed that lower charges on Li/Na ions correlate with good cycling behavior, so the low charges of Co, Li, and O are in nice accord with the superb performance of $LiCoO_2$. Alternative cathode materials with trivalent Fe^{3+} are given by $LiFePO_4F$, $NaFePO_4F$, and Na_2FePO_4F; see again Figure 4.27. $LiFePO_4F$ does not fully deintercalate Li because this would lead to inaccessible Fe^{4+}. Although there is an additional negative charge by F^-, this gets compensated by iron, and the Li charge remains about the same (Ertural et al., 2022). Note that an alternative way to arrive at transition-metal charges would be given by directly integrating the atomic magnetic

---

17 Such **ad hoc** correction (bridging the gap to analytic many-electron Mott–Hubbard-style approaches) of the undercorrelated standard DFT method is often done for solids, even if DFT then looks a bit more empirical. If that fails, however, better theories (e.g., dynamical mean-field theory, or GW) are at hand (Biermann & Lichtenstein, 2017).

moments, and then one arrives at charges in astonishingly good accord with Löwdin charges.

For typical anode materials such as graphitic carbon, the archetypes LiC_6 and LiC_{12} are shown in Figure 4.28 (left), truly layered structures consisting of pristine graphite with intercalated lithium, the decisive difference being that LiC_{12} contains Li only in every second graphene layer. Here, the meta-GGA functional SCAN (strongly constrained and appropriately normed semilocal density functional) is particularly useful. Li exhibits exactly the same wave function-derived Löwdin charge of +0.83 in both compounds, and C is charged −0.14 and −0.07. Alternative charge assignments based on densities, not wave functions, scatter between −0.05 and −0.23 for LiC_6 and −0.02 and −0.14 for LiC_{12}, due to simple numerical inaccuracies. Since the Li charge is rather moderate, does this explain the good intercalating behavior and does it also explain that Na does not intercalate? Covalency may also play a role (Lenchuk et al., 2019) but this is easily checked by a quantum-chemical analysis also given in Figure 4.28 (right).

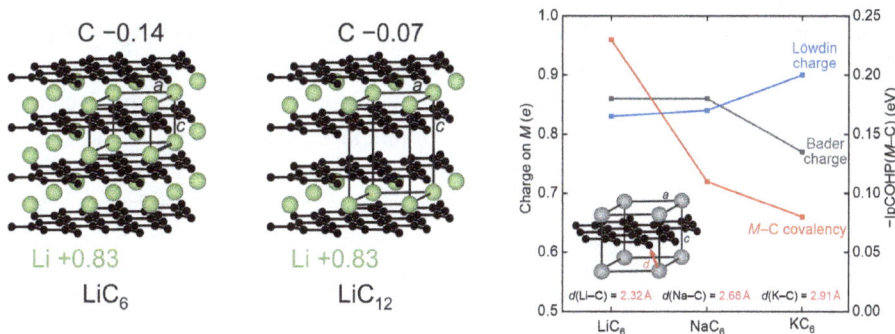

Figure 4.28: The crystal structures of LiC_6 and LiC_{12} (left) and theoretical Löwdin charges and integrated COHP for the combinations Li–C, Na–C, and K–C (right).

The integrated projected COHP values for Li–C, Na–C, and K–C show that covalency decreases in that order, in good accord with chemical knowledge, so LiC_6 stands out at being **most covalent**. Likewise, the Löwdin charges reveal that the K phase is the most ionic, in accord with ENs. So, the competition between ionicity and covalency is obvious for such anode materials (Ertural et al., 2022). It is interesting to note that density-based (Bader) charges lead nowhere because they suggest the K phase to be **less** ionic, in clear conflict with ENs. And such weird charges for K, clearly nonchemical, have already been spotted in the literature (Lenchuk et al., 2019), too, around +0.75, obviously pointless.

The superior behavior of wave function-derived charges, even for lithium, is particularly impressive when calculated for **amorphous** carbon species, depicted in Figure 4.29. Such carbon nanomaterials with substantial degrees of disorder are promising, and they can be generated by machine-learning-based interatomic potentials (Huang et al., 2019).

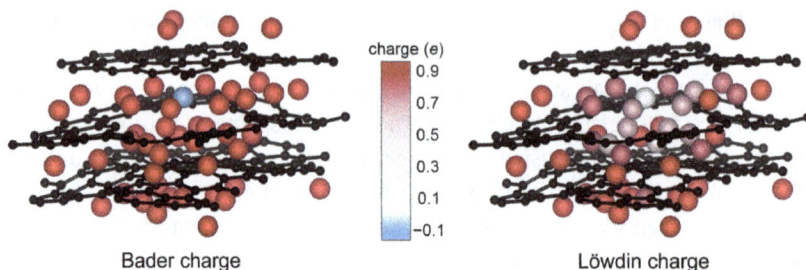

Figure 4.29: Calculated charge distribution for Li atoms intercalated into an amorphous graphitic carbon model as calculated by density-derived Bader (left) and wave function-derived Löwdin (right) charges.

The density-based charge calculation (left) is rather slow, and it also yields a chemically nonintuitive result, once again, namely, a **negative** charge on one central Li atom in a more electronegative matrix of carbon, chemically impossible. The wave function-derived Löwdin charges (right) are calculated far more quickly, and in the pore center there is an almost metallic (uncharged) Li ion, from which other Li ions with increasing charges are clustered around (Nelson et al., 2020). It has been argued that pores in disordered carbons can be filled with metallic-like clusters, and evidence of a (partial) charge transfer has been obtained, e.g., from operando NMR measurements on Li anodes (Letellier et al., 2003), yielding approximate charges of +0.66 and +0.1 in such environments.

The importance of atomic charge (balanced against covalency) in battery materials is so strong that virtually all kinds of physicochemical phenomena are being interpreted by that notion. Generally, lower ionic charges are expected to nicely correlate with an improved intercalation, and such hypothesis is theoretically checked quite easily, namely, by comparing charges and activation barriers side by side, as depicted in Figure 4.30, showing that the idea does **not** work, however. To do so, the activation barriers are derived from an initial and a transition state of the compound under question, a standard procedure.

Astonishingly enough, there is **no** convincing correlation of that kind found in whatever cathodic or anodic material. That being recognized, the intercalation and diffusion process must be driven by numerous other factors such as ionic charge **plus** size effects (by the ion, by vacancies, and by the surrounding sites), the individual migration pathway, possibly volume change as well, or even other structural effects. Nonetheless, charge is important.

In order not to frustrate the busy computational scientist, let us nonetheless try to theoretically **design** a novel battery material, something that is not tried very often but which will become routine in the future. First of all, let us keep all the charges rather small (and covalency strong), which is not a bad starting point. That translates into **not** using transition metals to begin with (some, like Co, are heavy, rare, and expensive, by the way), and we are also not focusing on oxides because of their −2 anionic charge in the ionic limit. Keeping in mind the high covalency of Li incorporated

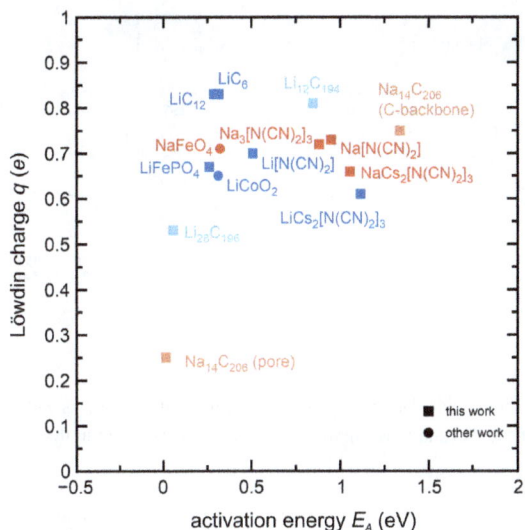

Figure 4.30: Comparison of calculated Löwdin charges of the migrating ion (Li or Na) with the activation energy in various electrochemically active compounds. Literature values in circles, and those by Ertural et al. (2022) in squares.

into a matrix of nitrogen atoms such as in Li_3N, the superb electrolyte, N instead of O is another good starting point but the large charge of the N^{3-} nitride anion must be lowered, to be done by going over to a complex anion such as NCN^{2-} which we have already witnessed in the conversion reaction at the very beginning of this chapter and which was introduced in the chapter before.

But there is even a better alternative, the dicyanamide anion, $[N\equiv C–N–C\equiv N]^- = N(CN)_2^-$, a kind of "fusion" of two NCN^{2-} anions, with an even lowered anionic charge of only −1. This complex anion has a boomerang shape, with a kink at the central N atom; chargewise, the dicyanamide (called "dca" from now on) resembles a simple halide anion such as Cl^-. And when there is a simple LiCl phase, there must also be a simple Li(dca) phase, and there is. And there are many more examples of such materials, also including other alkali metals, even ternary variants. A quick Löwdin charge calculation yields $Li^{+0.70}(dca)^{-0.70}$, to be compared with $Li^{+0.60}Cl^{-0.60}$, indicating that the **entire** dca anion is not very far from Cl^-, even though Li–N must be more covalent than Li–Cl. And the Li charge is similar to what is found in an anode material such as LiC_6, actually slightly lower. Figure 4.31 shows the structure of Li(dca) and the charge assignment (Ertural et al., 2022).[18]

---

18 The Bader charges are a bit unrealistic in this case, approximately +1.6 and −1.3 for C and N, respectively.

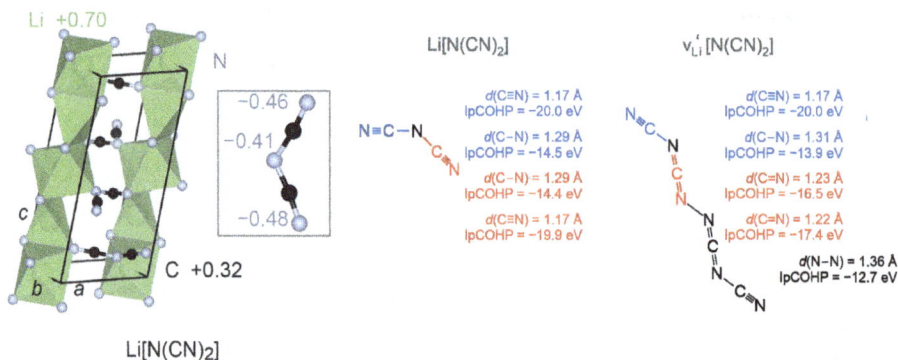

Li +0.70

N

-0.46

-0.41

-0.48

C +0.32

c

b a

Li[N(CN)2]

Li[N(CN)₂]

$d(C≡N) = 1.17$ Å
IpCOHP = −20.0 eV

N≡C−N

$d(C−N) = 1.29$ Å
IpCOHP = −14.5 eV

$d(C−N) = 1.29$ Å
IpCOHP = −14.4 eV

$d(C≡N) = 1.17$ Å
IpCOHP = −19.9 eV

$v_{Li}^{'}[N(CN)_2]$

$d(C≡N) = 1.17$ Å
IpCOHP = −20.0 eV

N≡C−N

$d(C−N) = 1.31$ Å
IpCOHP = −13.9 eV

$d(C=N) = 1.23$ Å
IpCOHP = −16.5 eV

$d(C=N) = 1.22$ Å
IpCOHP = −17.4 eV

$d(N−N) = 1.36$ Å
IpCOHP = −12.7 eV

Figure 4.31: The crystal structure of lithium dicyanamide, Li[N(CN)₂], and its Löwdin charges (left) as well as integrated COHP chemical-bonding data of the dicyanamide anion in Li[N(CN)₂] and its dimer in the delithiated $v_{Li}^{'}[N(CN)_2]$ (right) with a Li vacancy.

The C atoms just hold the dca unit together and do not interact any further, but the charge difference for the N atoms mirrors their functionality, either at the center (less charged) or at the terminal position (higher charged). Now, one can imagine what is going to happen if one Li atom is taken out from Li(dca), so we are considering theoretical delithiation in whatever battery. Because the activation barrier shown in Figure 4.30 is not too far off from LiCoO₂, this idea might work. Hence, we would end up with a structure dubbed □(dca) where □ stands for the Li atom being gone through delithiation; chemically, the new compound would simply read C₂N₃ but then a problem shows up: with 2 × 4 + 3 × 5 = 23 valence electrons, there must be one **unpaired** electron, so C₂N₃ is a **radical** which should show enormous reactivity. And this is true, because further structural relaxation gets rid of imaginary modes by making two radicals come together, forming an N–N bond and thereby pairing the two unpaired electrons, as also shown in Figure 4.31 (right).

The Lewis sketches of the radical C₂N₃ and its dimer C₄N₆ (not showing imaginary phonons any longer, this material is stable) are accompanied by the integrated COHP data, thereby directly evidencing their energetic character in terms of C–N bonding. But even without such quantum-chemical calculation, the driving force is pretty clear. Tabulated bond-dissociation energies for a homolytic bond breaking in HC≡N, H₂C= NH, H₃C–NH₂, and H₂N–NH₂ yield that the energy gain for cleaving a C≡N triple and a C–N single bond together with forming two C=N double and a new N–N single bond is about 263 kJ mol⁻¹, so dimerization is energetically favored, getting rid of the radical character.

And then one may theoretically design a battery, first by repeating all the calculations with the same DFT **ansatz** (PBE + D3) for all the components. Second, the resulting electrochemical potential ΔE is approximated from the battery equation, that is, $\Delta H \approx \Delta G = -nF\Delta E$ (F is the Faraday constant, 1 mol of electrons). For example, a potential battery would be composed of Li(dca) and graphite, following

$$2\,Li\big[N(CN)_2\big]_{(s)} + 2\,C_{6(s)} \rightleftharpoons \big(\big[N(CN)_2\big]\big)_{2(s)} + 2\,LiC_{6(s)}$$

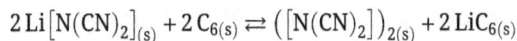

but the enthalpy is positive, so Li(dca) is not anodic in character but cathodic. Surprise! Hence, we assume reversibility and write the reaction as

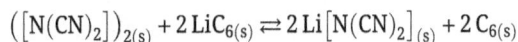

$$\big(\big[N(CN)_2\big]\big)_{2(s)} + 2\,LiC_{6(s)} \rightleftharpoons 2\,Li\big[N(CN)_2\big]_{(s)} + 2\,C_{6(s)}$$

and this yields an exothermic reaction enthalpy of about -391 kJ mol^{-1} which corresponds to a voltage of about $+4.1$ V, quite comparable to LiCoO$_2$ (Ertural et al., 2022).

Such dicyanamides are stable in acids and bases, and polymerization of Li(dca) only occurs above 300 °C, indicating thermal stability. The density ratio to LiC$_6$ arrives at 0.7 for Li(dca), so this phase (and related dicyanamides) could also compete in terms of bulk properties with commercially used electrode materials. Indeed, there is some rumor that "organic" electrodes (which here means **devoid of metals**, not devoid of C–C bonds because dca is inherently inorganic) are a promising alternative to overcome the sodium atom's intercalation issues or to avoid environmental or toxicity concerns. Since Li(dca) has a theoretical voltage of $+4.1$ V and a theoretical capacitance of 367.2 mAh g^{-1}, it should be fine. Admittedly, the computationally estimated activation energy for Li ion migration of Li(dca) is still a little larger than the one of LiCoO$_2$, so more research is needed, in particular experiments. And we do not know yet whether or not this battery design is reasonable at all, my very intention was to show how it could proceed **in principle**, by looking at charges and covalent bonding, too.

To summarize those issues as regards charges and chemical bonding in such battery materials, here are a few bullet points:

- Battery materials and also solid-state electrolytes show a fine balance of covalency versus ionicity, needed to both assure structural stability and also mobility for the (ionic) species moving between cathodic and anionic phases.
- Both intercalation and conversion reactions exist, and the aforementioned point is valid for both.
- Sufficient/insufficient covalency makes us understand the ability/inability of Li/Na to intercalate into graphite; charge calculations based on wave functions instead of density are not only much faster, they make more chemical and physical sense.
- Interpreting each and every effect (such as activation barriers) only by charges is a gross oversimplification, and there are a plethora of other factors.
- It is possible to computationally design novel cathode materials devoid of transition metals but containing complex N-based anions and arrive at very reasonable physical parameters, to be substantiated by experiments.

4.4 Molecular crystals, hydrogen, and other secondary bonding

When looking into standard textbooks of inorganic crystal chemistry, one is tempted to assume that most crystals are composed of **ions**, the eternal oversimplification, and the first chapter dealing with oxides, halides, sulfides, etc. would be quite typical for such notion. Admittedly, there are also extended covalently bonded crystals – diamond or silicon being prototype representatives – but there is more to come. Known from our daily experience, there are also so-called **molecular** crystals, that is, crystals composed of molecules which – when the crystals melt – do not change their chemical composition, for a very good reason, going back to the relative strengths of chemical bonding, most easily explained from a simple example. A material such as $BaTiO_3$ only exists as a solid and as a melt, too, because when trying to evaporate $BaTiO_3$ it will simply decompose. Sure, Ba^{2+} and Ti^{4+} may exist in the gas phase but O^{2-} will not, opposed to what is written sometimes (Appendix C). Hence, the very strong chemical-bonding forces used to keep refractory $BaTiO_3$ together in the condensed phase cannot survive, so to speak, when trying to transfer it into the gaseous state.

How different is the situation for molecular crystals, say, a crystal made up from simple water molecules, H_2O. This very molecular compound of paramount importance exists as ice, regular water, and water vapor, as we all know. So, the **identity** of the underlying molecular species, H_2O, **defines** the compound, and it does not change as a function of the physical state (solid, liquid, and gaseous). That can only be true if and only if the chemical forces within the molecule (mostly covalent bonds) are **much** stronger than the intermolecular forces (dispersion, hydrogen bonding, etc.), and we will now talk about that. Let us start with an element, though.

A crystal made up of only carbon atoms is not restricted to the diamond or graphite polymorphs[19] but one can also generate a carbon crystal made up of fullerene, the icosahedral C_{60} molecule resembling a soccer ball in which 60 C atoms form interconnected pentagons and hexagons. Let us discuss the electronic structure of an isolated C_{60} molecule that has been calculated by plane-wave DFT using a supercell such as to allow for enough space between the molecules; it is depicted in Figure 4.32 (left).

In the molecule, one finds a shorter and a longer C–C distance, 1.40 and 1.45 Å, naively interpreted as "double" and "single" bonds. In the DOS, there are plenty of molecular-orbital-like "spikes," so we are truly witnessing an isolated molecular species. If these are combined into a regular solid (not shown), the levels broaden slightly, because the molecules interact weakly with each other, mostly through much weaker dispersion forces.

The COHP comparison immediately shows the difference for the "single" and "double" bonds: for the longer "single" bond, the bonding situation is almost optimized, the integrated COHP being −5.9 eV, and it seems that two additional **bonding**

---

19 When talking about an elemental polymorph, the term **allotrope** is a little more appropriate.

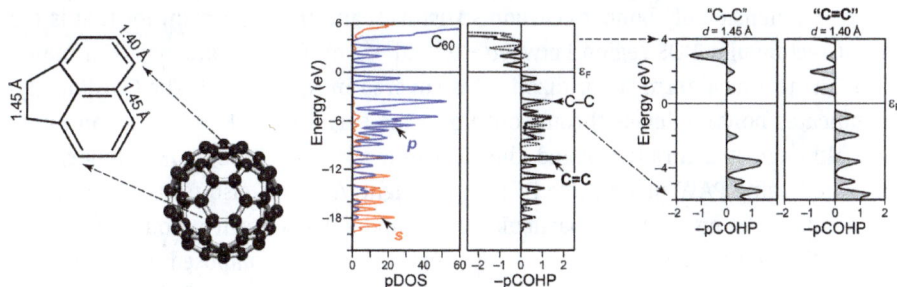

Figure 4.32: Geometric details of a C_{60} molecule (left) as well as (projected) DOS and COHP chemical-bonding analysis of short and long C–C interactions (right). Because these data have been projected from plane waves, DOS and COHP go under pDOS and pCOHP.

levels **above** the Fermi level might be fillable but this assumption is wrong. These levels are antibonding with respect to the shorter "double" bond which is also stronger (ICOHP = −6.4 eV), so there is some bond hierarchy at work. The stronger bond dictates the structural details, alternatively expressed, in harmony with the electron filling level (Maintz et al., 2013). As alluded to before, there is not much change upon condensation, so it is not at all surprising that C_{60} also exists as a solid, not only as an isolated molecule.

Another (infinite) molecule, now one-dimensional, is given by a carbon nanotube (of which there are many), depicted in Figure 4.33, together with the electronic structure of the essentially two-dimensional carbon polymorph graphite and its bonding analysis.

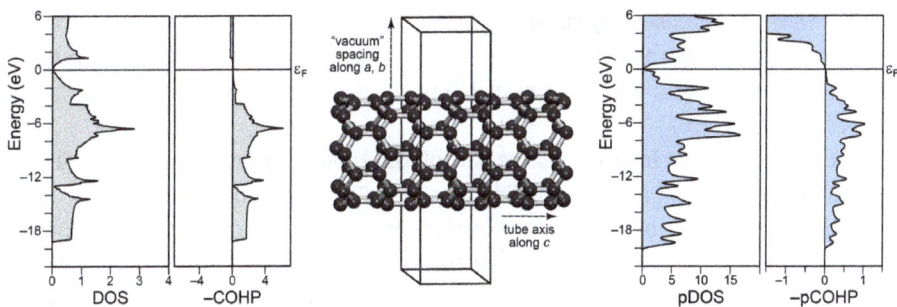

Figure 4.33: The electronic structure (DOS and COHP) of graphite (left), the structural details of a supercell calculation of an infinite carbon nanotube (center), and the electronic structure of the latter (right).

For a proper comparison, let us first look at the DOS and chemical-bonding analysis of graphite, left part of Figure 4.33, this time calculated with an all-electron local-orbital basis (LMTO theory). Graphite serves as the "archetype" of "sp^2"-hybridized (i.e., two-dimensionally bonded) carbon structures, and the fingerprint for that is the almost rectangular DOS region between −19 and −12 eV for such systems (Hoffmann, 2013). The technical term "hybridization" is covered in Appendix H. The COHP analysis indicates bonding levels throughout up to the Fermi level. For the carbon nanotube, although structurally isolated, the electronic structure (DOS) looks similar, now calculated using PAW plane-wave DFT on the right. And this also translates into the C–C bonds in the nanotube, in particular the nonbonding character around the Fermi level.[20] Whether we look at an infinite but curved molecule composed of hexagonally arranged carbon atoms or at a condensed carbon form made up of planar sheets of such carbon atoms, the result is **almost** the same, the reason being that one graphene sheet is only weakly perturbed by the other sheets (Maintz et al., 2013) since only a little dispersion between the two is involved.

Significantly stronger interactions are found for molecular crystals in which hydrogen bridging bonds are present, the aforementioned ice crystal being the prototype example, simple sugar would be another one. The hydrogen bridging bond, as already mentioned in Chapter 3 and the opening remark of this chapter, is a complicated mélange of three different bonding types. There is a covalent contribution, there is also some charge transfer (hence, ionicity), and there are dispersion forces at work, so sheer convenience has brought the chemists to call it bonding type no. 5, nicely covered in the literature (Pimentel & McClellan, 1960; Jeffrey, 1997) due to its paramount importance, mostly for biochemistry. From a more practical point of view in terms of defining our topic, a $D–H{\cdots}A$ hydrogen bond appears if a polar-covalent $D^{\delta-}–H^{\delta+}$ bond (i.e., including **both** covalency and ionicity) interacts with a polar atom $A^{\delta-}$, and the designators D and A symbolize the donor and acceptor atoms. In a C–H$\cdots$O combination, donor and acceptor would be different but they can also be the same such as in ice with O–H$\cdots$O. And one may classify hydrogen bonds based on their energy since the community more or less agrees that there are (mostly linear) strong (>63 kJ mol^{-1}), moderate (17–63 kJ mol^{-1}), and weak (<17 kJ mol^{-1}) types of H bonding (Steiner, 2002), simply **defined** that way.

Because dispersion interactions are part of **any** H bond by definition, DFT is probably not the best method for that bonding type. In DFT, everything is about the density, and when the density goes to zero (dispersion interactions do **not** require in-between electron densities), there is an obvious problem.[21] Hence, wave function-based calculations are required, often on molecular entities, as depicted in Figure 4.34, depicting clusters of water molecules.

---

20 It is not at all surprising that the C–C bond strength as quantified by COHP in the many allotropes of carbon – infinite in number (Hoffmann et al., 2016) – is a fine total-energy descriptor (Görne & Dronskowski, 2019) and may help in identifying other (likely) allotropic targets, yet to be synthesized.

21 In principle, yes, but not necessarily in practice. For example, the simple LDA "overbinds" such that dispersion at bonding distances is "modeled" by a fortunate error cancelation, unlike in the GGA case where an explicit correction needs to be used. For any functional, one simply needs to try.

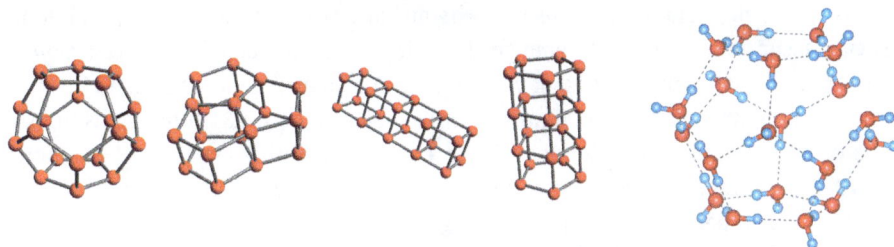

Figure 4.34: Four different kinds of water clusters of the composition $(H_2O)_{20}$ (left, O atoms depicted) and the lowest-energy dodecahedral $(H_2O)_{20}$ isomer in which two H bonds have been broken (right).

Here is a standard example of how to proceed: for a given entity of 20 connected water molecules dubbed $(H_2O)_{20}$, a number of high-symmetry configurations are thinkable, for example, dodecahedron, edge-sharing pentagonal prisms, fused cubes, and face-sharing pentagonal prisms (Figure 4.34). For the highest symmetry variant (the dodecahedron), even using the "ice rules" for distributing the H atoms in the cluster (Bernal & Fowler, 1933) still arrives at a very sizeable number of isomers, 30,026 in total, and the energy differences are also small, to make things worse. To make things disastrous, allowing for just one "broken" H bond will already yield 443,112 isomers, and 2 "broken" H bonds will generate an astonishing number of 2,772,313 isomers all of which must be correctly calculated. That **is** possible using wave function-based methods if one uses strictly localized **geminals** (i.e., two-electron functions, not one-electron functions = orbitals) and some semiempirical parameterization of the integrals. Despite the huge amount of computational data for a relatively small cluster, one may then identify low-energy candidates (see Figure 4.34, right) and also arrive at qualitative results: for example, the relative size of the $(H_2O)_{20}$ cluster nicely scales with the average Mulliken charge on the H donor atoms (highlighting ionicity), so the polarization of the H bonds is highly important (Tokmachev et al., 2010), thanks to the electronegative O atoms involved.

For extended solids, calculations of that kind look practically impossible to conduct, and the importance of dispersion interactions **between** molecules is still a problem. Because we cannot wait for the *ab initio* theory (in its purest sense) of hydrogen bonding in crystalline solids,[22] one may think of a compromise: covalency and ionicity pose no problem for DFT, so one simply adds a somewhat empirical (classical) dispersion correction to the exchange-correlation energy (Grimme et al., 2011) and calls it a "+D" method. We will come back to that below but, for the moment being, let us try a rather economic DFT functional first, the PBE well-known from solid-state theory, in order to see what can be accomplished even **without** any correction.

---

22 To quote a world-famous inorganic chemist: "Anyone who wants to harvest in his lifetime cannot wait for the *ab initio* theory of weather" (von Schnering, 1981).

As a first step, one might consider whether or not DFT done this way is able to provide accurate (or accurately enough) H positions in molecular solids, even though van der Waals (vdW) or any kind of weak interactions may pose a problem, to be compared with neutron-diffraction data. So, let us look at Figure 4.35 showing us the results of such computational endeavor. Because determining accurate H positions from X-ray diffraction, even single-crystal data, is rather difficult, plane-wave DFT could provide a helping hand in optimizing C–H, N–H, O–H, and B–H units found in whatever molecular crystals, the first step of analysis. One simply needs to calculate the Hellmann–Feynman forces and shift the H atoms as long as the forces deviate from zero, an automatic procedure.

The case of paracetamol (*N*-acetyl-*p*-aminophenol), an analgesic drug, is representative, and one of the polymorphs of the molecule has been clarified from single-crystal X-ray diffraction, but most of the C–H, N–H, and O–H bonds come out too short, the typical problem, in particular for the terminal methyl group. DFT, however, easily corrects the problem, in harmony with neutron data, even when DFT is **not** supported by a dispersion correction (within the crystal, not in an isolated gas-phase molecule), an astonishing result.[23] And it is perfectly okay to **selectively** optimize the H positions, not only because one stays closer to the experiment but the accuracy is sufficient already.

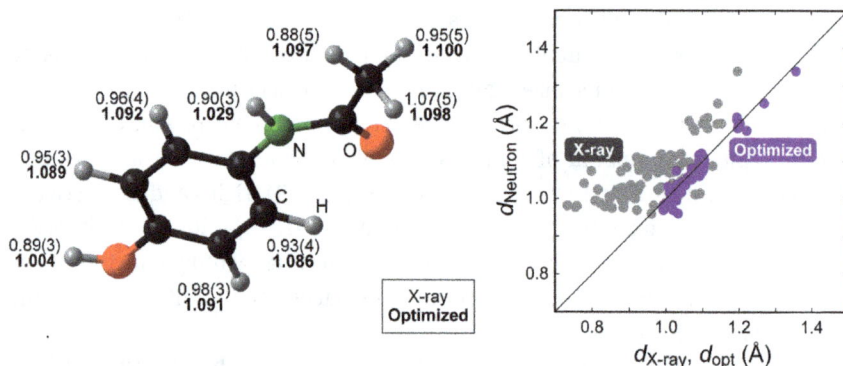

Figure 4.35: Molecular structure of paracetamol with individual bond lengths (Å) in the crystal structure determined via XRD and standard deviations in parentheses together with DFT-optimized ones in bold (left) as well as a comparison of a large number of bond-length data involving H atoms as found from XRD, from neutron diffraction and optimized from theory (right).

From a more general perspective, a large number of both XRD- and also neutron-determined bond-length data may be compared with theoretically derived ones

---

23 To put this into perspective one might say that DFT is an approximate theory but X-ray structure determination is an approximate theoretical deduction from approximate experimental data, empirically corrected with theoretical predictions, in particular as regards H positions.

(Figure 4.35, right), and this also supports the ability of highly economic DFT to come up with H positions (and, hence, "good" bonds to H) which almost yield neutron accuracy, so to speak (Deringer et al., 2012a). That being said, it is high time to check whether the H-bonding in a complicated molecular crystal structure can be quantified, for example, the crystal structure of guanidine which indeed shows a rather complex array of H bonds. Its complexity may be the reason why guanidine was structurally clarified so belatedly (Yamada et al., 2009) since the molecule is astonishingly difficult to crystallize. Figure 4.36 shows a typical sketch of what can be done computationally.

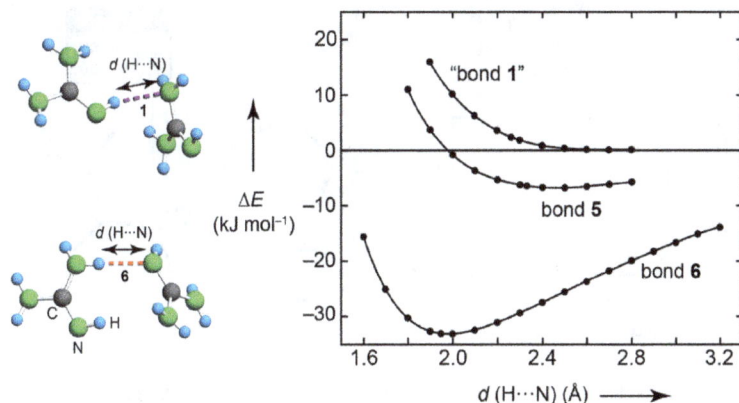

Figure 4.36: Sketches of molecular dimers in the crystal structure of guanidine and their energetic behavior (for bond **6** and "bond **1**" only) upon enlarging the distance of an individual H bond.

A dimer made up of two molecules including one particular H bond is put into a super-cell, and then the two molecular units are pulled apart from each other, quantified by the energetic change visible from Figure 4.36. Interestingly enough, not all of the eight different H bonds designated as such based on geometrical means are similarly strong. Most are moderately strong, two are weaker, and one is **not** an H bond (with no minimum in the energy–distance diagram, "bond **1**"). And by looking at the entire crystal structure, one may also quantify their cooperativity, that is, what all of them **gain** in energy compared to the simple sum of all the energetic parts. Taken as a whole, cooperativity and anticooperativity (i.e., the sum may be **smaller** than expected) stabilize the entire H bonding network by 14%. Admittedly, those H bonds in guanidine are strong **enough** to not require additional dispersion corrections, a lucky coincidence (Hoepfner et al., 2012). A subsequent single-crystal structure analysis based on neutron data supported all those structural details provided before using DFT (Sawinski et al., 2013).

Now that the energetics of hydrogen bonding and the important contribution of ionicity have become transparent, at least semiquantitatively, how about covalency? There are traditional crystallographic methods such as PIXEL to strive for that (Gavezzotti, 2008; Dunitz & Gavezzotti, 2012) but electronic-structure theory together with quantum-chemical analysis should do as well. To do so, the crystal structure taken by the compound *N,N*-dimethylbiguanidinium bis(hydrogensquarate) is a fine example, in particular because it has been determined using both high-resolution X-ray **and** neutron diffraction data. Exactly the same approach as used before can then be used to not only optimize the H positions but also to project out the chemical bonding as done using COHP, based on PAW plane-wave DFT and the PBE functional. The result is shown in Figure 4.37.

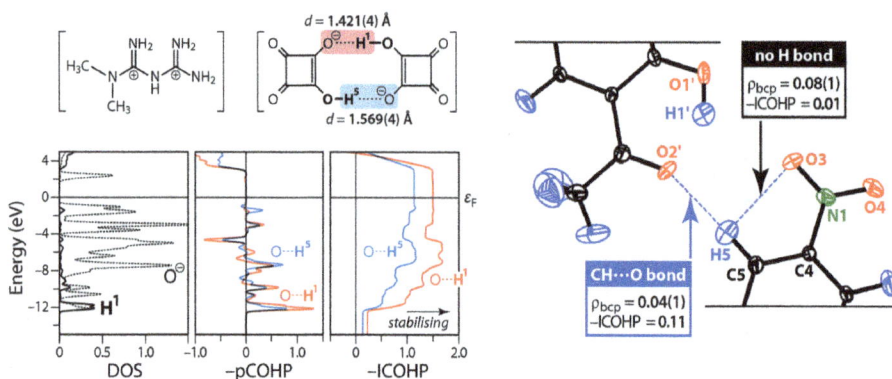

Figure 4.37: Structural sketch of dimerized hydrogensquarate anions together with the electronic structure (DOS) and chemical-bonding analysis (projected COHP) of the short and long O–H bonds (left) as well as H-bonded fragment from the crystal structure of 1-(2-hydroxy-5-nitrophenyl)ethanone (right).

The crystal is "molecular," as visible from the MO-like spikes in the DOS. There are shorter (1.42 Å) and longer (1.57 Å) hydrogen bonds, and both are nicely separated in the projected COHP plot, mostly bonding but also a little antibonding (at around −5 eV). When integrated up to the highest level, the short H bond comes out stronger (ICOHP) than the long bond, with covalent stabilizations of 149 and 104 kJ mol^{-1}, a good estimate of the real bond energy. Let us emphasize one more time that these effects are **strong**, we can live without dispersion corrections. Likewise, it is rather trivial to separate real H bonds from computational artefacts by other approaches, as depicted in the right part of Figure 4.37. Besides a real intermolecular H bond between O2′ and H5 not only visible by COHP but also from the bond critical point of Bader's atoms in molecules (AIM) theory, there is also a short **intramolecular** contact (between O3 and H5) which COHP clearly determines as nonbonding although the bond-critical-point criterion would classify it as an H bond, incorrectly so. The latter AIM theory only relies on the charge density and its partitioning, so it does not contain the essential **phase information** of whatever involved orbitals, the decisive step in capturing the essence of covalent interactions. In this respect, an orbital-based method such as COHP does an excellent job (Deringer et al., 2014a).

That being said, it does not come as a surprise that even the abundant biopolymer chitin stabilizing the exoskeleton of insects and giving structure to plants may be analyzed in terms of its intricate H-bonding network connecting the polymer strands. Here, the H-bonding interactions cover a larger region in terms of energies, so dispersion-corrected DFT (see below) needs to be carried out for simulating the crystal network and lower-dimensional fragments (Deringer et al., 2016).

A moment reflection reveals that such "secondary" bonding interactions as found for the hydrogen bond involving a donor–acceptor three-atom configuration dubbed $D–X{\cdots}A$ not only exists when X = H. In fact, the crystallographers have long realized, by carefully observing unusual bond-length phenomena in crystalline materials, that something similar also happens for halogen atoms, so the term **halogen bond** was coined. The first observation of that kind goes back to the year 1863, in fact (Guthrie, 1863). And then **chalcogen bonding** was identified, followed by **pnictogen bonding**. Let us briefly summarize, with a simple sketch, what is meant by the term "halogen bond," to start with the simplest example resembling the hydrogen bond, and Figure 4.38 a) provides the proper illustration.

Using modern language, the term "halogen bond" (XB) coined in molecular chemistry designates relatively[24] weak bonding interactions of a halogen atom that is otherwise covalently bonded and whose local electronic structure corresponds to an already filled noble-gas shell. The halogen atom possesses a σ-hole (i.e., a region of positive electrostatic potential, ESP) to which a suitable partner with an outer negative ESP may be electrostatically attracted; once again, the importance of an ionic contribution is clear from the very beginning. An analogous definition is valid for the chalcogen and pnictogen bonds, of course, nothing changes.

To not only quantify these secondary interactions, in particular as regards their cooperativity (which became transparent for guanidine and chitin already), plane-wave DFT including a decent exchange-correlation functional such as PBE in addition to a "dispersion correction" is the right choice, as alluded to already. The latter correction needs to model the so-called London dispersion forces (1930), which turn elemental bromine and iodine into a liquid or solid-state material under standard conditions, respectively, unlike elemental fluorine or chlorine. Although the electron density between, say, iodine molecules drops to zero – a tough problem for DFT which is based on the density – there are "nonlocal" forces out of nothing (a consequence of Heisenberg's uncertainty principle) which decay with the sixth power of the interatomic distance (for long distances), hence London's early formula

$$E_{\text{London}} \sim -C_6/R^6$$

is self-explanatory; one "only" needs the correct C_6 coefficient in the simplest approach for correction. As this is not the right place to discuss the richness of the

---

24 The meaning of the word **relatively** depends on the point of view, of course.

(mostly weak) non-covalent interactions and their theories, we refer to the literature (Grimme et al., 2016; Otero de la Roza & DiLabio, 2017) and merely mention three DFT trades in coping with the problem, successfully so, but depending on the problem. One either adds a more or less empirical dispersion correction (Grimme et al., 2011) to the current density functional, as said before,[25] or tries to directly derive a nonempirical correction from the underlying DFT-based density (Tkatchenko & Scheffler, 2009) or utilizes entire functionals specifically made to model vdW interactions (Lee et al., 2010).

For characteristic systems, it is rather straightforward to compare all those pnictogen, chalcogen, and halogen interactions and find out about the trends, depicted in Figure 4.38.

A dimer of I–C≡N molecules experiences an interaction energy of about 23 kJ mol^{-1} which grows to 28 kJ mol^{-1} for three molecules and eventually to 41 kJ mol^{-1} for the infinite chain, a perfect example (79% boost) of cooperativity, the sum being more than its parts. And this effect not only exists for iodine (Figure 4.38 b), but is also visible for the lighter halogen atoms, albeit to a lesser extent, even for fluorine. Clearly, the polarizability of the halogen atom (maximum for iodine) plays an important role. And it is instructive to compare such theoretical results with experimental high-resolution electron densities and also Raman frequencies (Wang et al., 2019).

Likewise, a similar trend can also be seen (Figure 4.38 c) for the chalcogen bonds and the pnictogen bonds, once again showing increasing cooperativity as the atoms get heavier, the chalcogen bonds of $Te(CN)_2$ and the pnictogen bonds of $Sb(CN)_3$ being most extraordinary (George et al., 2014). From an alternative perspective, the tendency of the heavier homologues to engage in extra bonding **beyond** the noble-gas shell is in perfect harmony with their ability to not follow the octet rule (Appendix A); the heavier the atom, the less it is compliant with the bonding of the first long period (N, O, F in the above cases). Anticipating things to come, we will talk about multicenter bonding (also called hyperbonding) in the final chapter because it all belongs together.

Before we touch on the most extreme case of tetrel bonding (yes, it also exists, at least it has been defined in that context), a word of caution may be needed. While the energetic trends shown above find their correspondences in experimental observations as regards stabilities of molecular crystals, there are also prominent examples of spectacular **failure** in the description of related, yet even simpler molecular crystals. Nobody really wants to talk about that to not anger the gods of DFT, but let us do it here at least quietly. Take elemental iodine, I_2, for example, the failure for bromine is almost as bad.

If one optimizes the ground-state crystal structure of I_2 with a decent density functional with or without dispersion correction, one arrives at the well-known molecular structure showing intramolecular I–I distances of about 2.8 Å (the I–I single

---

25 For example, this approach has allowed to computationally characterize various phosphorus allotropes whose energetics heavily depend on vdW interactions (Bachhuber et al., 2014).

(a)

halogen
bond
$-R \cdots \overline{|X} - C \equiv N|$

halogen
atom

(b)

$I-C\equiv N \cdots I-C\equiv N$
$\Delta E_{int} = -22.8$ kJ/mol

$I-C\equiv N \cdots I-C\equiv N \cdots I-C\equiv N$
$\Delta E_{int} = -28.2$ kJ/mol

$\cdots \equiv N \cdots I-C\equiv N \cdots I-C\equiv N \cdots I-C\equiv N \cdots I-C$
$\Delta E_{int} = -40.8$ kJ/mol

(c)

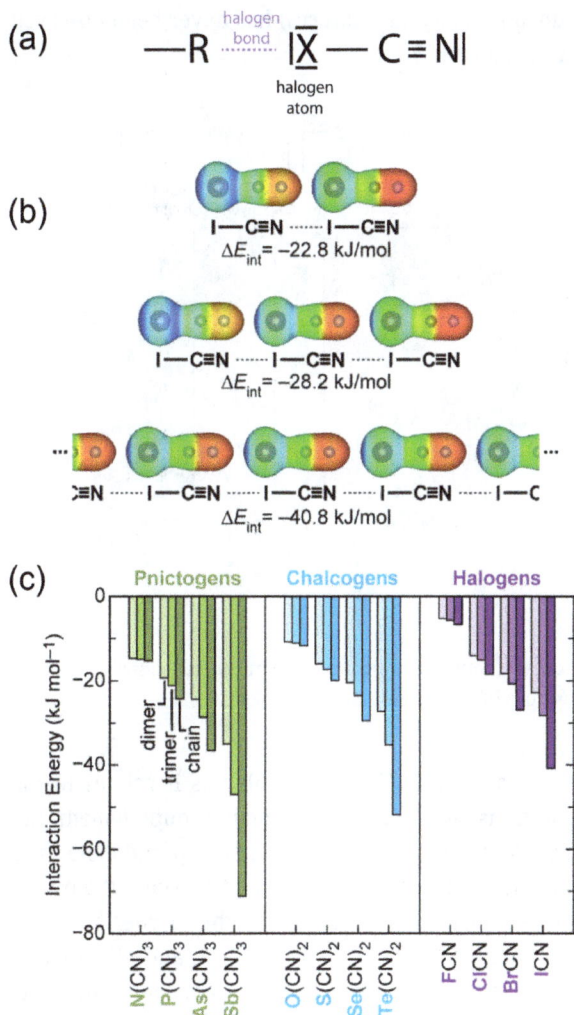

Figure 4.38: (a) Molecular icon of a typical halogen bond, (b) halogen-bonding cooperativity as reflected from charge-density isosurface (0.05 e A^{-3}) plots and interaction energies as well as (c) cooperativity diagrams for various pnictogen, chalcogen, and halogen bonds expressed via interaction energies as a function of the main quantum number.

bond) and much wider intermolecular distances of 3.4 Å, corresponding to the vdW distance between molecules, in harmony with experiment. This structure is **unstable**, however, as reflected from imaginary phonon modes, so a small kick lets the atoms move and the structure change into another structure which is even **lower** in energy and dynamically **stable**, and this structure is characterized by an infinite zigzag chain with equal I–I distances of 2.9 Å. This structure, somewhat resembling the result of an

inverse Peierls distortion (from unequal to equal distances) has never been observed experimentally, and it is depicted in Figure 4.39.

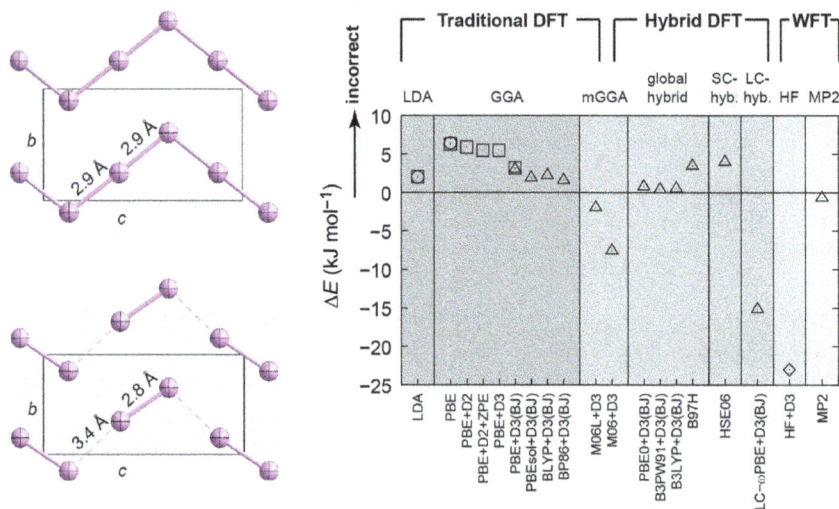

Figure 4.39: The incorrect (top left) and correct (bottom left) ground-state structure of crystalline iodine as well as the energetic performance of various DFT functionals for this problem.

This result is a sheer disaster for standard DFT, in particular as it relates to the ground state, not excited-state questions, and to a "simple" main-group element, not some "correlated" species such as a $3d$ atom. Neither basis sets nor pseudopotentials nor faulty algorithms are to blame, the blame goes to the core of DFT and the operative functional.[26] Admittedly, two meta-GGA functionals get it right, as does one long-range corrected hybrid functional, and also periodic Hartree–Fock and MP2 theory, wave functions are clearly superior in this respect. We still do not know the exact amount of incorrectness (George et al., 2015) but the tendency of DFT to "delocalize" and equilibrate all the I–I distances, incorrectly so, is profound. Please keep in mind that caution is advised in terms of energies for halogen bonds, and a satisfying structural agreement with experiment may mean **nothing** as to the correctness for the energies. How many incorrect theoretical structures still need to be identified?

That being said, let us finally look at tetrel bonding as displayed in Figure 4.40. The "tetrels" are members of main-group IV (or group 14), and we will deal with the carbon, germanium, and tin tetracyanides, $Tr(CN)_4$, serving as characteristic examples.

There are at least two ways to quantify tetrel bonding, given a prior plane-wave DFT calculation and an exchange-correlation meta-functional such as M06L which

---

26 It is **not** a dispersion problem; otherwise, the blame (and the cure) would be obvious.

ICOHP(Tr=C) = –923 kJ mol^{-1}
ICOHP(Tr=Ge) = –586 kJ mol^{-1}
ICOHP(Tr=Sn) = –402 kJ mol^{-1}

ICOHP

ΔE_{int}(Tr=C) = –467 kJ mol^{-1}
ΔE_{int}(Tr=Ge) = –382 kJ mol^{-1}
ΔE_{int}(Tr=Sn) = –220 kJ mol^{-1}

ΔE_{int}

... N≡C▬Tr ··· N≡C—Tr ···   N≡C▬Tr ··· N≡C—Tr ···

ICOHP(Tr=C) = 0 kJ mol^{-1}
ICOHP(Tr=Ge) = –45 kJ mol^{-1}
ICOHP(Tr=Sn) = –288 kJ mol^{-1}

ΔE_{int}(Tr=C) = –15 kJ mol^{-1}
ΔE_{int}(Tr=Ge) = –34 kJ mol^{-1}
ΔE_{int}(Tr=Sn) = –130 kJ mol^{-1}

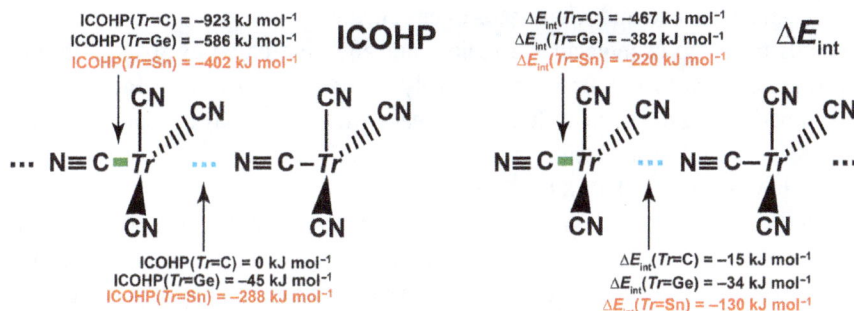

Figure 4.40: Tetrel interactions in a chain of Tr(CN)$_4$ seen from the perspective of integrated crystal orbital Hamilton populations (ICOHP, left) and interaction energies (right).

was behaving well in the prior bromine and iodine cases. First, one may simply integrate the COHP of the various bonds, both the "regular" (to the left of the central Tr atom, in green) and "tetrel" ones (to the right, dotted in blue), simply because that measures their contributions to the band-structure energy (the sum of the Kohn–Sham eigenvalues). The comparison is depicted in Figure 4.40 (left). Upon going down main-group IV, the tetrel bond strength grows rapidly, and it is 0 for C (no covalency, only weak dispersion) but –288 kJ mol^{-1} for Sn in terms of ICOHP. At the same time, ICOHP of the regular bond weakens from –923 kJ mol^{-1} (C) to –402 kJ mol^{-1} (Sn). So what is gained in tetrel bonding strength gets lost (to a smaller degree) in the regular bond.[27]

Second, exactly the same tendency can be derived from calculated interaction energies per bond between the parts of the chains, shown in Figure 4.40 (right). Here one simply cuts the different monomers apart from each other but keeps their geometrical details intact, impossible in practice but easily possible in theory. Upon going down, from C to Sn, the tetrel bond strengthens and the regular bond weakens. For Sn, the tetrel bond reaches ca. 59% of the regular single bond, an astonishing amount, in the range of very strong hydrogen bonds.

While none of the two approaches is physically "right," they do serve our understanding. Note that ICOHPs represent an arbitrary partitioning (in orbital and energy space) of the band-structure energy while interaction energies correspond to a likewise arbitrary partitioning (in real space, without relaxation), so both are true *Gedankenexperiments*. Not only does covalency increase while going down main-group IV in the above examples, cooperativity shows the same trend. In certain crystal structures it can be shown that up to 70% of the crystal stabilization is, in fact, caused by the tetrel bonds (George & Dronskowski, 2017).

---

27 Again, these numbers relate to the lowering (by covalency) of the band-structure energy only, and measurable bond energies are **not** meant by that.

What can be learned and what needs to be concluded?

- Crystals made up of molecules are electronically dominated by strong internal (intramolecular) forces almost exclusively coming from covalent interactions. When packed together, the influence of weaker intermolecular (dispersion) interactions between the molecules has almost no influence on the electronic structures, they remain "spiky" in the DOS.
- The hydrogen bond is an important exception; this complex mélange of covalency, ionicity, and dispersion interactions may become rather strong. Its quantitative calculation calls for wave function-based calculations, at least in principle.
- If hydrogen bonds are not too weak, a decent exchange-correlation functional will excel in yielding H positions of "neutron accuracy," and it will also help to semiquantitatively characterize the strength of hydrogen bonding by the covalent contribution; if the numbers are small, however, dispersion corrections are a must for DFT.
- The so-called halogen, chalcogen, pnictogen, and tetrel bondings are also secondary bonding types, strongly related to hydrogen bonding, and usually dubbed "σ-hole" interactions. DFT failures are known to exist for these, however.
- The aforementioned bondings increase in strength when moving down the main groups; the heavier the atom, the higher the tendency of octet widening and secondary bonding, and the higher the amount of cooperativity.

4.5 Zintl phases, metals, itinerant magnets, and steels

We are going to deal with metals, intermetallic compounds (Pöttgen & Johrendt, 2019) and further derivatives of the latter in this chapter but, in order to properly connect to the former chapters in which significant charge transfer and, hence, ionicity was important, let us first think about the influence of ENs when metals make up compounds purely consisting of metals. Does that sound weird? Not necessarily! A moment reflection reveals that an intermetallic compound composed of, say, elemental Rb and Cs should lead to a phase dubbed $Rb_{1-x}Cs_x$, a metallic solid solution (both existing as solid and liquid phases), given that the ENs and atomic sizes of Rb and Cs are comparable. That is correct, and exactly the same arguments let us also correctly assume the existence of an intermetallic compound dubbed $Cu_{1-x}Au_x$ which likewise behaves as a metal. If we were to combine Cs and Au, however, we would end up with a semiconductor CsAu in which the charge distribution corresponds to Cs^+Au^-, thanks to the very electronegative and "relativistic" Au (somewhat similar to Br and I, see Appendix G) and the "electropositive" Cs. That seems surprising at first sight but similar effects can be quite common in intermetallic compounds, for example, in the case of Cs_2Pt which is also a dark-red semiconductor (Karpov et al., 2003), a **salt** made up of two metals.

The aforementioned transition (or collapse?) of metallic bonding into ionic bonding given a sufficiently large EN difference was first recognized by Zintl in the 1930s (Schäfer et al., 1973; Kauzlarich, 1996), an intellectual concept particularly helpful and valid for main-group metals, and nowadays a quick electronic-structure calculation on binary phases such as NaTl and KTl will corroborate these very early ideas. For illustration, Figure 4.41 shows structure and ionic bonding in those two compounds.

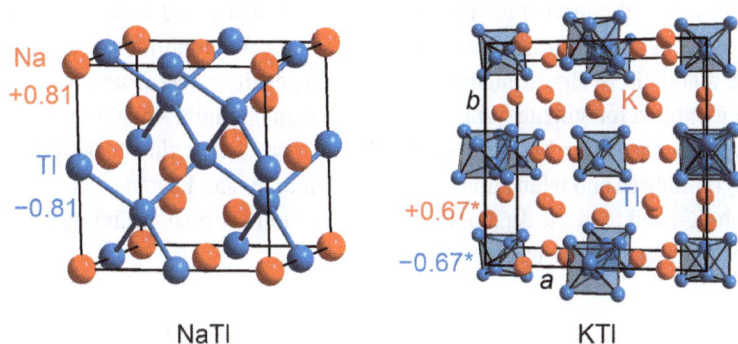

Figure 4.41: Crystal structures of the two Zintl phases NaTl and KTl with quantum-chemically calculated Mulliken charges. Asterisks designate averaged values.

Within NaTl, the charges are seemingly so large (±0.81) that one may think of a ionically bonded semiconductor – a bold but correct assumption – and a charge transfer of practically one entire electron from Na to Tl. Hence, Tl from main-group III then formally becomes a main-group IV atom such as carbon, and then the anionic substructure formed by the Tl atoms (namely the diamond network) makes perfect sense. For KTl, the charge transfer is somewhat smaller, insufficient to generate the diamond substructure, so nature lets six Tl atoms form an octahedron formally charged –6; the radii also have an influence. Many more examples exist: KGe may be formally considered K^+Ge^-, so Ge turns into a main-group V element such as P; hence, the structure contains Ge_4^{4-} tetrahedra resembling P_4 molecules. Likewise, CaSi may be formulated as $Ca^{2+}Si^{2-}$, making Si a kind of main-group VI element such as Se, and it does not surprise us that CaSi contains kinked Si chains resembling Se helices. Such **a posteriori** affirmations of the Zintl concept are fun to carry out given proper wave function-derived charge calculations, and a plethora of examples exist for that purpose (Ertural et al., 2019).

Admittedly, the Zintl concept is a gross oversimplification but it serves its purpose in the hands of a well-trained chemist, useful whenever the EN differences are truly large. Quite another intermetallics concept is the one by Laves, focusing on the individual atomic radii and how they fit each other to allow for a good packing, $MgCu_2$ being a fine example (Simon, 1983). And then there are the so-called Hume-Rothery (or brass) phases which are – so the idea – controlled by the valence-electron concentration and how that fits a certain structure type. On purpose, we will continue to discuss selected examples of bonding in intermetallics, in particular because we want to highlight new findings of practical interest for engineers; for a larger chemistry-focused overview, there is an excellent review article (Lee & Fredrickson, 2017).

Between the Zintl and Hume-Rothery phases, however, the region of **polar intermetallics** pops up, materials that are poorer in electrons than the Zintl phases. For these, simple electron counting (in particular, with idealized charges) does not make sense at all, but how should one then navigate in terms of structure and bonding? As a typical example, we look at the mineral **stützite**, $Ag_{5-x}Te_3$, a compound formed by the metal Ag and the metalloid (= almost metallic) Te. A homogeneity range from $x = -0.25$ to $+1.44$ has been identified for stützite. Which x would give an optimum stability? This is difficult to answer, in particular because the hexagonal crystal structure looks quite messy, with $(Te^-)_2$ dumbbells and tellurium nets of honeycomb (h) and Kagome[28] (K) type, stacked as –h–K–h–K–, and then the Ag atoms are distributed in-between, including positional disorder for both Ag and Te (see Figure 4.42).

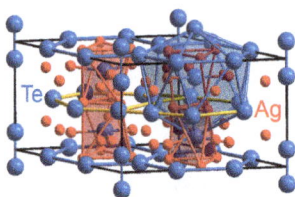

Figure 4.42: Simplified crystal structure of stützite with the Kagome-type Te nets in yellow.

In terms of compositions, various possibilities such as "$Ag_{32}Te_{21}$" or "$Ag_{34}Te_{21}$" or "$Ag_{36}Te_{21}$" come to mind, but the small absolute EN difference between Ag (4.44 eV) and Te (5.49 eV) forbids to make a little charge calculus using Ag^+ and Te^- or Te^{2-} because nothing fits. Admittedly, one might just forget about charges and look at the quantum-mechanical formation enthalpies which render "$Ag_{34}Te_{21}$" as most stable, followed by "$Ag_{32}Te_{21}$," then followed by "$Ag_{36}Te_{21}$" but that is unsatisfactory, at least for chemists. So, Mulliken charges are calculated from the PAW plane-wave electronic structure and the GGA, and this yields $(Ag^{+0.37})_{32}(Te^{-0.66})_{13}([Te^{-0.40}]_2)_4$ for "$Ag_{32}Te_{21}$," $(Ag^{+0.36})_{34}(Te^{-0.68})_{13}$ $([Te^{-0.36}]_2)_4$ for "$Ag_{34}Te_{21}$," and $(Ag^{+0.33})_{36}(Te^{-0.64})_{13}([Te^{-0.42}]_2)_4$ for "$Ag_{36}Te_{21}$." The **trend** is important, not the second decimal place. So, there are lesser charged Te atoms forming dumbbells (like I_2) and higher charged Te atoms as isolated "anions" (like [Xe]). And due to these charges, the Madelung energy of "$Ag_{34}Te_{21}$" is $-4,304$ kJ mol^{-1}, more stable than "$Ag_{32}Te_{21}$" with $-4,138$ kJ mol^{-1} and "$Ag_{36}Te_{21}$" with $-3,873$ kJ mol^{-1}. One needs to confess that judging such polar intermetallics electrostatically is a simplified approach but it does yield chemical information (Ertural et al., 2019).

---

28 The Japanese word "Kagome" is derived from a traditional woven bamboo pattern with a trihexagonal tiling.

Alternatively, quantifying chemical bonding in intermetallics by focusing on COHP is also possible, and it is particularly useful when the charges are even smaller (Steinberg & Dronskowski, 2018), like in many polar intermetallics, the "black sheep" of intermetallic compounds because they are so difficult to describe using classical means. Another, even more difficult example would be given by $CsCe_2Ag_3Te_5$, an intermetallic telluride presumably comprising ionic as well as strong (polar) mixed-metal bonds, certainly not to be handled by any Zintl reasoning. Its electronic structure based on GGA calculations is shown in Figure 4.43. The underlying crystal structure is likewise complicated, composed of Te tunnels which encompass Cs, Ce, and Ag atoms, so complicated that we skip showing it; it is not needed.

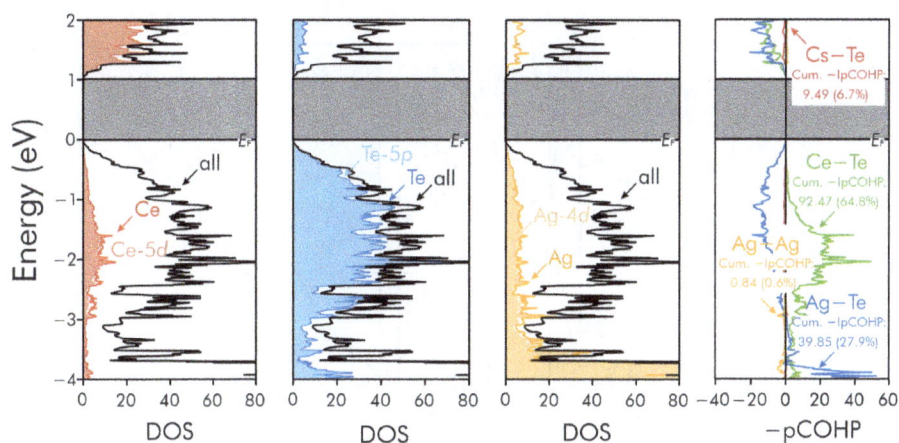

Figure 4.43: Densities of states and projected COHP chemical-bonding analysis of $CsCe_2Ag_3Te_5$.

The Löwdin charges arrive at +0.67 for Cs, +0.57 for Ce, +0.44 for Ag, and −0.63 for Te, so ionicity plays a role but not a major role. Instead, it seems more reasonable to focus on the covalency, easily quantified by the cumulated ICOHP values (projected from plane waves), also depicted in Figure 4.43, and here it helps to compare the ICOHP percentage contributions. That being said, the valence band is heavily dominated by Te, the mostly "anionic" species, but Ce (the $5d$ orbitals) and Ag ($4d$) also mix in. More quantitatively, Ag–Ag comprises only 0.6% of the covalent interactions because it corresponds to a $d^{10}-d^{10}$ interaction,[29] preceded by Cs–Te (6.7%), then by Ag–Te (27.9%), and finally by Ce–Te (64.8%), the latter atomic pair forming the strongest covalent bonds in this intermetallic compound (Eickmeier et al., 2020). Clearly

---

29 For the transition metals, talking about $(ds)^{10}-(ds)^{10}$ interactions would be more fitting because the ns orbital is energetically a little above $(n-1)d$.

the Ce–Te bond rules this phase, and chemical-bonding theory based on DFT helps to distinguish important from less important interactions.

Fortunately, the chemical bonding of simple metals is much simpler, let us concentrate on the basics. Take Na in the body-centered crystal structure in which each Na has eight nearest neighbors but only **one** 3s valence electron for bonding, so seven out of eight neighbors are not bonded in a two-center two-electron bond unless the electron gets socialized over all those. And this is how covalency mutates into metallicity, too few electrons for too many atoms, an extreme case of electron poverty, the "free electron" (or Sommerfeld) model of solid-state physics (Ashcroft & Mermin, 1976). The corresponding crystal structure, band structure, and DOS are displayed in Figure 4.44.

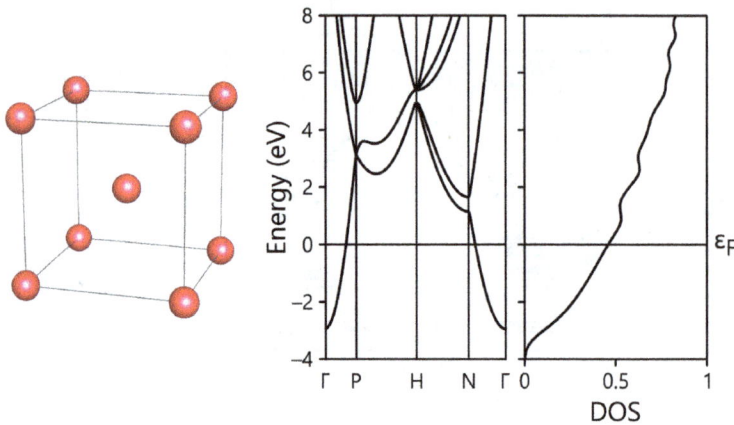

Figure 4.44: Crystal structure, band structure, and DOS of *bcc*-Na.

The band structure shows several bands but only one (3s) is partially occupied, and its dispersion is strong, meaning that the band must be isotropic and also extremely delocalized in energy and space, so we have recovered the Sommerfeld model by PAW plane-wave theory. Given the low electron count, there will be practically no antibonding levels (try calculating yourself using LOBSTER). The Na–Na distance, the only free parameter in the body-centered cubic (*bcc*) structure type, must have adjusted to the electron count, and it did. And such a scenario is also found in other "simple" metals such as Ca or In or Pb, at least semiquantitatively.

For the transition metals, the situation is quite different, simply due to the more nodal nature of the *d* orbitals involved. Take Fe in the same *bcc* structure type, for example. To keep things transparent, we run a density-functional calculation using the simplest functional (the LDA), and we also treat α- and β-spins (or spin-up and spin-down) the same, that is, there is no spin polarization to begin with. The fundamental and likewise astonishing result is presented in Figure 4.45.

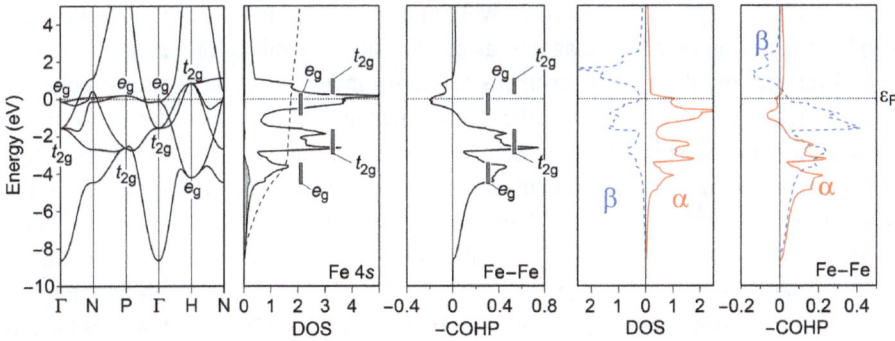

Figure 4.45: Band structure, DOS, and COHP of *bcc*-Fe using a non-spin-polarized LDA calculation (left) as well as spin-polarized (α-spin in red and β-spin in blue) DOS and COHP (right).

The bands go up and down, and it is easy to distinguish the one band with the strongest dispersion (which stems from 4*s* similar to Na 3*s* before) from the five other (the more contracted five 3*d*). The DOS is almost entirely 3*d* in character, and it also shows the characteristic "three-peak" shape made up by the e_g and t_{2g} levels, so typical for all *bcc* transition metals. The chemical bonding, however, yields, in addition to low-lying bonding levels, a likewise **occupied antibonding** level just at the Fermi level. That cannot be true, and it is not a fault of the functional.[30] One would expect a structural distortion (in the sense of the Peierls effect) to happen but *bcc*-Fe stays *bcc*-Fe, and the structure does not change.[31] Nonetheless, the onset of spin polarization (Figure 4.45 right) changes everything: the majority spin channel (α) goes down, the minority spins (β) move up, a consequence of the exchange hole (Dronskowski, 2005) which makes like spins (α–α or β–β) avoid each other. Given equal amounts of α and β (say, two and two), there is no change, but for three α- and one β-electrons, there is a massive effect. Since the α-electrons cannot shield each other very well, they see a higher nuclear charge and move toward the nucleus, whereas the better shielded β-electron moves up.

And all that translates into chemical bonding, because the formerly antibonding level almost vanishes by the α-spins going down and the β-spins going up, so there results some bonding stabilization, on the order of 5%, and **that** is the chemical reason for *bcc*-Fe becoming ferromagnetic, namely, the optimization of its chemical bonding, thereby generating a magnetic moment of about 2.2 electrons. Analogous but

---

30 While the LDA is unable to correctly quantify the energetics of the Fe allotropes, this is no problem for the GGA. Both GGA and LDA, however, show the same effect as regards the occupied Fe–Fe antibonding level. It is **fundamental** in terms of VEC and structure and **not** related to the accuracy of the exchange-correlation functional used.

31 The body-centered tetragonal crystal structure of indium, however, may be understood as a Peierls-distorted variant of the *fcc* crystal structure (Häussermann et al., 1999).

smaller effects exist for Co and Ni as well, and only these two transition metals are also known as itinerant ferromagnets, as our physicist friends call them. It also needs to be said that the β-minority electrons are more involved in the chemical bonding than the majority α-spins because the β-spin orbitals (or spinors, in short) are more diffuse, another consequence of the exchange hole, and this explains why – in theory – spin-polarized metals have **wider** interatomic distances than non-spin-polarized ones (Landrum & Dronskowski, 1999; Landrum & Dronskowski, 2000). Likewise, such chemical theory of itinerant magnetism also succeeds in understanding itinerant antiferromagnetism such as in *bcc*-Cr for which the Fermi level cuts through nonbonding levels in the non-spin-polarized calculation (Decker et al., 2002b). And it allows for additional insight when looked at by an entire theory of spin polarization (Seo, 2017).

Because the effect is a function of the electron count and exchange splitting, it can also be observed for other itinerant systems, for example, an artificial layer of face-centered cubic (*fcc*)-Ru on *fcc*-Ag (Maintz et al., 2016). And the same arguments in favor of ferro- and antiferromagnetism can, in fact, be turned into a chemical recipe to make itinerant magnets **by design**, on purpose. This includes an early theoretical-synthetic study on ternary/quaternary metallic borides (Dronskowski et al., 2002) which was then elaborated to yield an entire series of such magnetic material.

For example, take the phase $Sc_2FeRh_5B_2$ which crystallizes tetragonally in an ordered variant of the so-called $[Ti_3Co_5B_2]$ structure type. $Sc_2FeRh_5B_2$ has a VEC of 65, because there are 2×3 (Sc) + 8 (Fe) + 5×9 (Rh) + 2×3 (B) = 65 electrons in total. In the non-spin-polarized calculation, the Fermi level goes through Fe–Fe antibonding levels, hence one correctly expects a ferromagnet. Even for the phase $Sc_2FeRh_3Ru_2B_2$ for which the VEC is $63 = 2 \times 3$ (Sc) + 8 (Fe) + 3×9 (Rh) + 2×8 (Ru) + 2×3 (B), the highest occupied levels are still antibonding, so this phase is also a ferromagnet. In contrast, the composition $Sc_2FeRu_5B_2$ with VEC = 60 is an antiferromagnet because this VEC corresponding to 2×3 (Sc) + 8 (Fe) + 5×8 (Ru) + 2×3 (B) lets the Fermi level lie in Fe–Fe nonbonding levels. The series covers the entire 60–65 range for the VEC, amazingly enough (Fokwa et al., 2007); although the origin of spin polarization (always the same Fe atom) stays identical, the VEC and the bonding character at the Fermi level determines everything. A wealth of different cases are found for complex intermetallics, too (Entel, 2017).

The interested reader may think that such elaborated solid-state chemical arguments, even when they lead to new itinerant magnets, look rather aloof, probably close to academic l'art-pour-l'art, but this could not be farther from the truth. The same train of thought including chemical-bonding analysis, if properly carried out, may directly be used for improving the most important structural material of all times, **steel**, and it may even contribute to the first-principles design of novel **types** of steel. Simply speaking, regular steel is a composite material mostly made up of iron with a very small carbon content to improve its mechanical properties (Holleman et al., 2007; Greenwood & Earnshaw, 1997), that is, strength and fracture resistance.

And there are many different types of steel which, in total, make steel probably also the most complicated structural material ever.

Quite recently, high-strength high-manganese steels became the very core of modern metal engineering, for a number of reasons, so *ab initio*-based modeling of the Fe–Mn system looked like a rational way to follow. To understand the important crystallographic phases in that realm, the key components of regular steel had to be analyzed first. For example, a ferromagnetic phase dubbed **cementite**, Fe_3C, is one of the most common and important phases in carbon steels. It crystallizes in the orthorhombic cohenite type whose unit cell is depicted in Figure 4.46 and in which the C atoms are trigonal-prismatically coordinated by Fe atoms sitting on 8*d* and 4*c* Wyckoff positions (Dierkes et al., 2013). For Mn_3C, the structure is the same but its properties are less well-known.

When investigating the phase diagram of $(Fe_{1-x}Mn_x)_3C$ by means of DFT, one finds good agreement with experiment in terms of structural, thermochemical, and also magnetic properties. For Mn_3C, a formerly unknown ferrimagnetic ground-state structure is found, however, in addition to an AFM structure, thereby highlighting the capability of Mn to induce **antiferromagnetism** in general, also in Mn-rich steels.

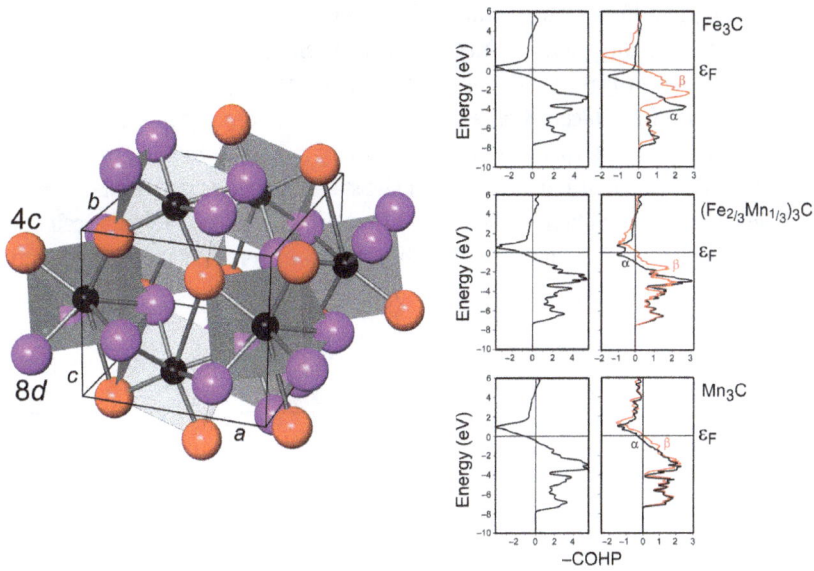

Figure 4.46: Look into the crystal structure of cementite (left) with metal atoms on 4*c* and 8*d* positions and C in black, and COHP chemical-bonding study of the various interactions assuming non-spin-polarized and polarized ground states (right).

Extended supercell calculations yield that the atomic distributions of the Fe and Mn atoms are **not** random-like but one finds three ordered regions within the phase range for which the key role is played by the 8*d* metal site forming differing magnetic layers as a function of composition. All effects in terms of magnetism, volume, and energy are easily explainable by COHP chemical-bonding analysis, also depicted in Figure 4.46 (right). In going from Fe_3C to Mn_3C via an intermediate phase, the Fermi level lowers, which results in an antibonding/nonbonding metal–metal character; hence, one finds a tendency to go for ferromagnetic (Fe_3C) or AFM/ferrimagnetic (Mn_3C) order. The higher the manganese content of intermediate phases, the more stable they turn out, thanks to the lowered electron count and, thus, fewer antibonding states at the Fermi level; note that Fe_3C is endothermic (positive enthalpy of formation) but Mn_3C exothermic (negative enthalpy of formation) (von Appen et al., 2010) and paramagnetic (Dierkes & Dronskowski, 2014).

Another intermetallic compound ubiquitous in manganese-rich steels is given by **austenite**, a phase of truly paramount importance. It is simply an *fcc* solid solution of Fe and Mn, and the interstitial C atoms are assumed to be randomly situated in the octahedral voids – but almost no short- or long-range experimental investigations exist, probably due to a lack of proper physical probes; it may be that there is Mn–C clustering (He et al., 2002), however.

The question whether or not austenite ordering of Fe and Mn atoms exists is straightforwardly answered by electronic-structure theory. A DFT study with a $Fe_{16}Mn_{16}C$ supercell containing 10 different structural conformations shows that there is a thermochemical driving force leading to manganese enrichment in the immediate proximity of the C atom. As depicted in Figure 4.47 (left), the relative energy lowers almost linearly from a pure Fe- to a pure Mn-coordinated C, the energy difference between CFe_6 and CMn_6 octahedral entities being ca. 33 kJ mol^{-1}, practically independent from the relative orientations of the atoms and also from the magnetic state. But where does that originate from? The result of a chemical-bonding analysis based on a larger $3 \times 3 \times 3$ supercell and 108 metal atoms, also shown in Figure 4.47 (right), lets us understand the effect found, and it couldn't be more surprising.

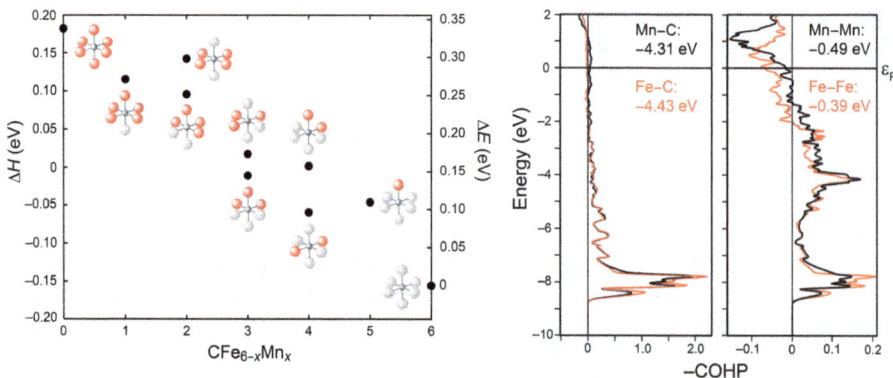

Figure 4.47: Different configurations and energetics of carbon bonded to Mn and Fe in an fcc solution of MnFe (left) and the underlying reason expressed by COHP bonding analysis (right).

To begin with, it turns out that the Fe–C bonding quantified by ICOHP is **stronger**, by about 3%, than the corresponding Mn–C bonding, an unexpected finding, but easily explained by reiterating that there is one more electron used by Fe to bond to C. So, the larger affinity of C toward Mn must have a different cause. The clue to understanding this affinity, however, is found in metal–metal bonds which **weaken** more upon C incorporation in an Fe_6 octahedron than in an Mn_6 octahedron. That is to say that the larger affinity of C toward Mn is an indirect effect. By swallowing C into an Fe_6 octahedron, the Fe–Fe bond length increases from 99% to 106% of the typical Fe–Fe bond length, overstretching it, thereby weakening orbital overlap. For Mn_6, carbon expands the Mn–Mn bond from 92% to a nearly perfect 99% of twice the Mn metal radius. The formerly compressed Mn–Mn bonds (which need to adjust to the Fe host) readily adapt to the carbon-induced widening and get stronger. Chemical bonding makes all the difference (von Appen & Dronskowski, 2011), but indirectly.

Given that the principles of atomic ordering and bonding proclivities are being understood, it is far easier to model the energetics of the entire Fe–Mn–C system based on the CALPHAD (CALculation of PHAse Diagrams) method, and one simply needs the best data for austenite (*fcc*), ferrite (*bcc*), the *hcp* phase, and also the aforementioned cementite phases, including metastable carbides such as Fe_3C, $Fe_{23}C_6$, Fe_5C_2, and Fe_7C_3 to reassess the literature data. And it works and provides an improvement, in particular in reproducing experimental liquidus and carbide equilibrium data at low temperatures (Djurovic et al., 2011). Likewise, the entire Fe–Mn phase diagram can be modeled from first-principles (see Figure 4.48), including nonmagnetic and AFM ordering, and compared with the CALPHAD approach, simply because the important phases are **understood** using chemical-bonding theory from the outset.

Clearly, such a phase diagram is done on purpose. All that is targeted at developing new high-manganese steels, with Mn contents of 15–30% and C mass contents of up to 1%, simply due to the fact that these twenty-first century steels exhibit high strength and exceptional plasticity due to extensive twin formation under mechanical load by the TWIP (= twinning-induced plasticity) effect, or via multiple martensitic transformations, the TRIP (= transformation-induced plasticity) effect. As given both by DFT and also empirical data (CALPHAD), Mn contents on the order of 20% induce close-packed structures and stabilize the austenite phase. So, the structure is no longer *bcc* as in pure Fe. Second, DFT also shows that these close-packed structures (either *hcp* or *fcc*) are energetically very close to each other, thus highlighting the possibility and importance of the TWIP effect. Third, it is obvious that the CALPHAD *bcc* curve for large Mn contents simply does not make sense since magnetism cannot be properly described classically (Lintzen et al., 2013).

Novel steels of that type have their problems, though, despite being less expensive and mechanically more promising than the existing ones, **hydrogen embrittlement** possibly leading to catastrophic materials failure being the correct engineering term. In close-packed structures, H accumulates and yields crack formation in strained regions. Octahedral sites are clearly preferred by hydrogen, and H also wants to be sur-

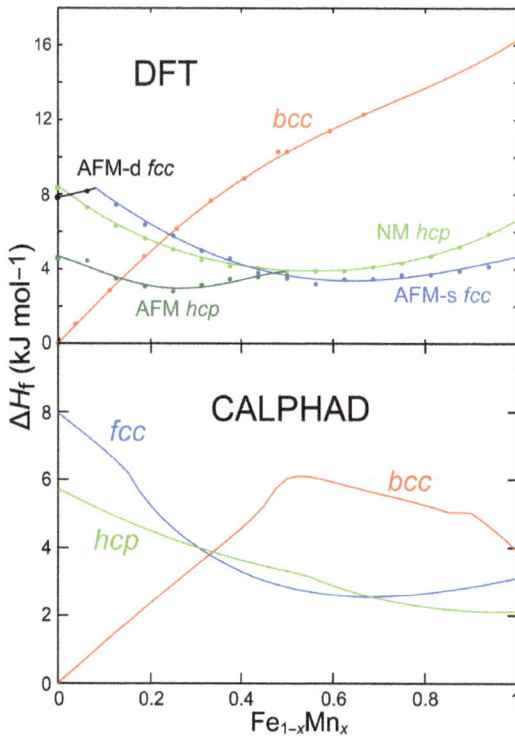

Figure 4.48: Phase diagram of the binary system Fe–Mn as given by density-functional theory (top) and the standard CALPHAD approach (bottom).

rounded by manganese, the chemical bonding being in control one more time. The situation resembles what was found for interstitial C, and the finding is nicely illustrated by the COHP diagrams depicted in Figure 4.49, again based on a $2 \times 2 \times 2$ supercell of composition $Fe_{16}Mn_{16}$ into which a single H is introduced but at different sites.

The COHPs of the Mn–H and Fe–H bonds look similar with a deep-lying single peak from the 1s H level plus smaller contributions between −8 and −5 eV. Regarding ICOHP, the Fe–H bond is marginally stronger (1.4%) than the one of Mn–H, −1.41 eV compared to −1.36 eV, as we already expect from the carbon case. When an empty Fe octahedron with all bonding levels optimized gets filled with H, however, the situation changes distinctly. The deep-lying hydrogen level at −10 eV also mixes into the iron levels and generates antibonding Fe–Fe levels close to ε_F indicative of bond weakening. In an Mn_6 octahedron, antiferromagnetism complicates things: Mn–Mn interactions between two antiparallel magnetic layers weaken somewhat when H is introduced but within one layer including a ferromagnetic Mn–Mn spin configuration, the incorporation of H does **not** weaken those Mn–Mn bonds. Hence, hydrogen diminishes all but one kind of metal–metal bonding, and so one will find chemical configurations that have more

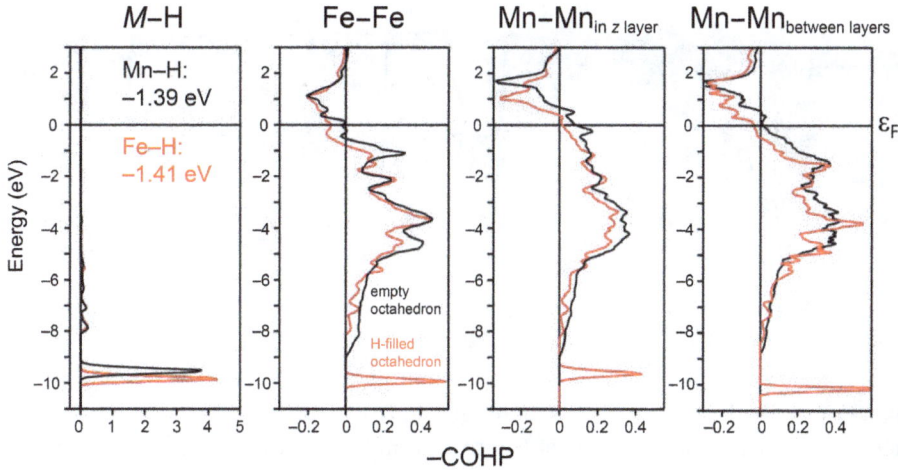

Figure 4.49: Chemical bonding of hydrogen in a binary FeMn model system as given by COHP and ICOHP using an all-electron DFT method (LMTO).

Mn–H contacts, in perfect agreement with theoretical (and experimental) hydrogen solution enthalpies (von Appen et al., 2014).

The next compositional step is given by Fe–Mn–Al–C austenitic steels with superior strain hardening behavior, and here the so-called κ-phase $(Fe,Mn)_3AlC$ plays an important role, its structure being depicted in Figure 4.50 (left), and geometrically it reminds us of the perovskite type (see Section 4.1). Spin-polarized DFT on an *fcc*-like $Fe_{106}Al_2C$ system reveals that Al has an even stronger impact on the ordering of the κ-phase than carbon (Song et al., 2015), so one finds an astonishingly strong Al–Al repulsion, and Al atoms tend to stay away from each other. Likewise, the most favorable C sites are all Al-free octahedra, and this effect of repulsive C and Al is even much stronger than the former Al–Al repulsion. As such, the κ-phase structural motif, clearly ordered, is easily explained **a posteriori**. And this is also corroborated by the phase-pure bulk synthesis of the κ-phase "Mn_3AlC," the exact composition being $Mn_{3.1}Al_{0.9}C$. Although C is fully occupied, there is a mixed occupation of the corner position both by Al and Mn, which leads to greater stability, as mirrored by DFT (Dierkes et al., 2017).

It now turns out that, in order to get rid of hydrogen embrittlement, "trapping" of the latter at some particular Wyckoff positions is a very worthwhile goal, and here the aforementioned κ-phase comes in handy since empirical data suggest that a small amount of Al lowers the tendency for such catastrophic failure. What happens on the atomic scale?

If one looks at a central H atom in an octahedral site in an *fcc*-Fe matrix, there is an energetic optimum if Al and H are second-nearest neighbors. In crystallographic terms, that situation matches an H atom having replaced the central C atom in the κ phase, hence corresponding to an Fe_3AlH κ-hydride phase; for reasons of brevity, we

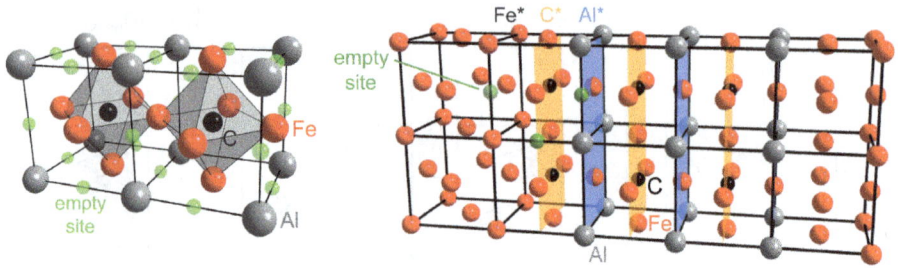

Figure 4.50: Supercell of the crystal structure of κ-Fe₃AlC (left) and supercell made up from the same κ-carbide and *fcc*-Fe (right).

ignore any H–H and C–H interactions that may also arise. A good computational model is shown in the right part of Figure 4.50, an intergrowth between the κ-carbide and *fcc*-Fe. All competing energetic trends taken together, the noncentral octahedral sites in the C-containing layer of an Fe*–C*–Al* block (i.e., the green sphere in the orange-colored layer designated C*) are, therefore, most attractive for H. At these particular sites, there are four C atoms in the second coordination sphere but only one Al atom in the first coordination sphere. Slightly less favorable are the octahedral sites in Fe layers adjacent to a C-terminated κ-carbide (green sphere in the Fe* layer) because then two C atoms in the second coordination sphere are available, but no Al atom in the first coordination sphere. These configurations would yield computationally designed H trapping at the interface between κ-carbide and the metallic matrix (Timmerscheidt et al., 2017).

Using similar techniques, it is also possible to theoretically model the formation of Mn–C short-range ordering in "X60Mn18" steel, based on a $3 \times 3 \times 3$ *fcc* 111-atom supercell with composition $Fe_{88}Mn_{20}C_3$ (Song et al., 2018), in nice agreement with additional neutron-diffraction data; the short-range ordering has a great influence on the mechanical properties of high-Mn steels. In real steels, C atoms will preferably agglomerate in Mn-rich regions, but keep a minimal C–C distance of at least more than 5 Å because there are repulsive C–C in addition to attractive Mn–C interactions.

Given such atomistic data, the Al–Mn–C system can also be thermodynamically evaluated by the CALPHAD method, in particular because DFT-based enthalpies help in modeling the κ-phase and elucidating general energetic trends (Tang et al., 2018). We recall that Mn_3AlC is better described as $Mn_{3.1}Al_{0.9}C$, so off-stoichiometric structures with antisite defects, either with Mn on Al positions or Al on Mn positions, were modeled and simulated. Full C occupancy in Al–Mn–C κ-carbides is energetically beneficial across the entire stoichiometric range, as also proven experimentally before. And all these *ab initio* results have meanwhile turned into **industrial products** because such steels are made worldwide for the automotive industry. You may drive them already, imagine that. There is nothing more practical than a good theory, here dealing with the chemical bonding in metallic matter.

Summarizing this very chapter dealing with intermetallic compounds and its derivatives, a few points come to mind:

- There are those intermetallics, in particular for main-group elements, where large EN differences let us consider them salt-like semiconducting (Zintl) phases, with immediate structural consequences; atomic charges play an important role.
- For less ionic phases such as polar intermetallics, ionicity also plays a role but the differing amounts of covalency – most easily quantified from ICOHP data – show which interactions are important and which are not.
- Chemical bonding in "simple" metals such as Na is likewise simple but things complicate quite a lot for the case of (magnetic) transition metals such as Fe. Here, the exchange hole has a massive influence, and becoming spin-polarized actually strengthens chemical bonding. A recipe for making ferro- and antiferromagnets is almost trivial to come up with.
- The chemical bonding also has a massive influence in the most important structural material, steel. All the stability and local structures scale with the chemical bonding, in particular metal–metal contacts, preferred carbon site, hydrogen embrittlement and trapping, short-range order of any kind.
- Given all the chemical-bonding and total-energy data, wisely generating thermochemical models of the main components of steel is eventually possible, thereby avoiding an infinite amount of useless calculations (or expensive and time-consuming experiments) to begin with.

4.6 Weird stuff from high pressure

Among the two thermodynamic variables, pressure and temperature, used in synthetic solid-state chemistry and materials science, pressure is clearly the less convenient one. Almost anybody can set the temperature to a (very) low number, either by liquid nitrogen or helium, and also (very) high temperatures can be reached by a standard oven or electric arc, hence chemical reactivity by temperature control is at our fingertips. To obtain the right pressure needed for chemical reactions, a greater experimental effort is needed, though. Nonetheless, at some instrumental places all required pressures can be realized, ranging from 1 MPa (about 10 atm) to more than 300 GPa (1 GPa = 10^3 MPa, about 10,000 atm) using a belt apparatus, (multi) anvil cells, and diamond-anvil cells. Even the pressure conditions of our Earth's interior (ca. 360 GPa at 5,000 °C) can be reached and surpassed, with important implications for synthetic solid-state chemistry (Huppertz et al., 2017).

Conventional high-pressure wisdom long before the advent of the GPa range can be summarized with two basic rules (but not laws) relating pressure with the periodic table as well as distances and coordination behavior. (1) Under pressure, an element aligns with the higher homologue. (2) Under pressure, the coordination number rises and the interatomic distances increase – which sounds counterintuitive only at first sight because the density increases **despite** bond lengthening.

The graphite-to-diamond transition at about 10 GPa and 3,000 °C is an example of the first rule because C then adopts the ground-state structure of Si, the diamond structure, so C aligns with Si. Likewise, the second rule is also fulfilled here because the coordination number increases from 3 to 4. Another example of the second rule is the transition from α-Sn to β-Sn at about 1 GPa when Sn changes from fourfold to distorted sixfold coordination. Likewise, Te increases its coordination number at higher pressure (Section 4.7). And there are many other examples of that kind, carefully and competently covered in the literature (Grochala et al., 2007; Zurek, 2017), even immensely chemical in spirit.

Other polymorphs that can be transformed into each other are the many high-pressure phases of germanium (Schwarz et al., 2008) showing four- and sixfold Ge coordination by Ge, fulfilling rule no. 2. And virtually every solid-state chemistry textbook covers the different stabilities of the SiO_2 polymorphs, namely, quartz, coesite, and stishovite, and these transformations occur at about 2 and 12 GPa, usually interpreted as affirmations of Pauling's third rule (see again Section 4.1) because condensation of the SiO_4 tetrahedra increases; in fact, that is a good point: compression leads to condensation, even valid for molecules. The normal molecular crystal of water contains H_2O molecules in which O bonds to two H atoms but it increases its coordination number to 4, above 60 GPa, in the ice X polymorph. Finally, we mention a more modern chemical example, the nitrides of silicon, germanium, and tin. The so-called spinel phase of Si_3N_4 with four- and sixfold coordination of Si by N can only be reached by high pressure (15 GPa) from either α- or β-Si_3N_4 with tetrahedral coordination, so this is rule no. 2 at work. The same high-pressure effect is found for Ge_3N_4 but tin nitride, Sn_3N_4, containing the higher homologue Sn already adopts the spinel structure in its ground state, so this confirms rule no. 1 (Scotti et al., 1999) since Si and Ge align with Sn.[32]

From a more physical point of view, there is the general idea, found in many solid-state physics textbooks, that pressure leads to an overall compression of matter, first by squeezing out "redundant" vdW space, which is easy. For extremely large pressures, everything – including hydrogen – will become metallic at a certain point.[33] In other words, extreme pressure should level all chemical differences. Whether pressure really leads to metallizing everything is a point worth discussing, so we will come back to that. And let us not forget – despite fundamental questions – that there is also a refresh-

---

32 I tried to find, without success, the **existing** phase Sn_3N_4 (reported in 1999, a true breakthrough in solid-state chemistry) in a prominent electronic-structure database specializing in materials genomics. It must be my fault.

33 The latest theoretical prediction taking into account many-body electronic correlation and quantum anharmonic motion results in a transition pressure of 577(4) GPa for **atomic** metallic and likewise superconducting hydrogen (Monacelli et al., 2023).

ingly practical aspect to high-pressure research since it often allows for making excessively hard materials (Riedel et al., 2017).

A more recent and fine example of pressure polymorphism in a very simple binary nitrogen-containing system is given by the high-pressure behavior of iron nitride, FeN. On purpose, I am concentrating on a more covalent system because then the changes in chemical bonding are less difficult to demonstrate. The story is easy to understand by looking at Figure 4.51.

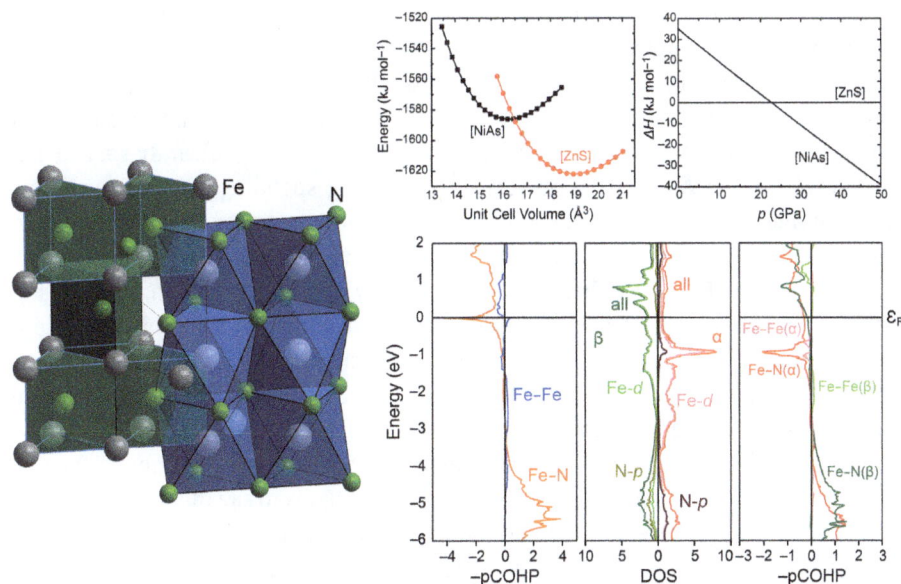

Figure 4.51: The high-pressure form of FeN crystallizing in the [NiAs] type (left) and energetic and bonding information of the same system based on electronic-structure theory (right).

FeN is part of the incredibly rich family of iron nitrides which are of paramount importance in materials research, used either as magnetic pigments or as surface protectants of steels. In addition, they are considered geoscientifically relevant because they may form part of the Earth's core (under enormous pressure). Under standard conditions, FeN should crystallize in the [ZnS] type (Dronskowski, 2005), somewhat depending on the N content. To arrive at a high-pressure FeN activated by high temperature, one simply needs to react Fe_2N with Fe in laser-heated diamond-anvil cells, accompanied by synchrotron X-ray diffraction and ^{57}Fe Mößbauer spectroscopy for characterization. At ca. 10 GPa, the new form of FeN manifests and adopts the [NiAs] structure type. This looks rather puzzling, because face-sharing FeN_6 octahedra are unknown in nitrides, and even the structure type is uncommon for that type of compound. As also seen in Figure 4.51 (right), the [NiAs] type has a smaller volume and a higher energy, so it is less stable; the theoretically predicted transition pressure can also be extracted.

As regards the electronic structure, it shows a high DOS at the Fermi level, and a non-spin-polarized bonding analysis detects strongly antibonding interactions just at the Fermi level, so we would expect spin polarization to occur. In harmony with the Mößbauer spectroscopic data, a magnetic ground state is present, and the ferromagnetic spin ordering is (slightly) lower in energy than the AFM one (Clark et al., 2017). So, upon transforming from [ZnS] to [NiAs], FeN increases its coordination number (rule no. 2) and loses in covalency because the loss in energetic stability goes back to weakened bonding as judged by COHP analysis.

There are other well-known cases of pressure polymorphism, reported for a long time already, for example, in sulfides of the lanthanide elements. Take samarium sulfide, SmS, for example. This phase crystallizes, almost boringly, in the [NaCl] type, and subjecting it to a moderate pressure of about 0.65 GPa leads to a transformation into an **analogous** [NaCl]-related arrangement[34] but with a significantly smaller lattice parameter, 5.88 Å instead of the former 5.97 Å, corresponding to a sudden volume drop, parallel to an increase in conductivity which corresponds to a semiconductor-to -metal transition. The scenario is a bit tricky to properly describe theoretically, in particular by using DFT, simply because Sm is a highly correlated $4f$ atom, so at least a Hubbard correction is needed. But we can arrive at a simple, intuitive model without any numerical calculation, and it goes like this: the change in structure and property (metallicity) relates to the oxidation states in SmS which will read Sm^{2+} and S^{2-} in the semiconducting phase. Nonetheless, the atomic **configurations** must differ, so they are close to $4f^6 5d^0$ in the ground state, the regular [NaCl] type resembling the CaS case, but $4f^5 5d^1$ in the high-pressure metallic phase. That is to say that one $4f$ electron is "activated" by pressure, and gets promoted into a $5d$ orbital with better overlap to its neighbors by a shorter lattice parameter, so the $5d^1$ level is smeared out (gets delocalized) in space and generates the metallic state (Batlogg et al., 1976). One may even formulate the metallic phase as $Sm^{3+}S^{2-}e^-$, thereby rationalizing the smaller lattice parameter by the smaller cationic radius of Sm^{3+}. Seemingly, pressure metallizes the material, the simple but compelling physical idea.

And there is even an allotropic transformation of a simple metallic element, the alkaline-earth strontium in this case, also known for a long time. The element Sr takes the *fcc* structure type at standard conditions but it changes to the *bcc* structure type at a pressure of about 3.5 GPa (Jayaraman et al., 1963). Figure 4.52 shows their electronic structures.

It seems strange, at first sight, that a close-packed structure (74% space filling) should change into a non-close-packed structure (68% filling), but the strontium atom **shrinks** at the same time, counteracting the worse space filling since the metallic radius

---

34 Note the wording, agreed upon by the picky crystallographers, meaning that the two [NaCl] structures are **similar** or **analogous** but not identical. Although both space groups and atomic positions are the same, the structures are only **isopointal** because the bonding differs. For being **isotypic** (= the structures are the same, lab jargon), the bonding should **also** be the same (Lima-de-Faria et al., 1990).

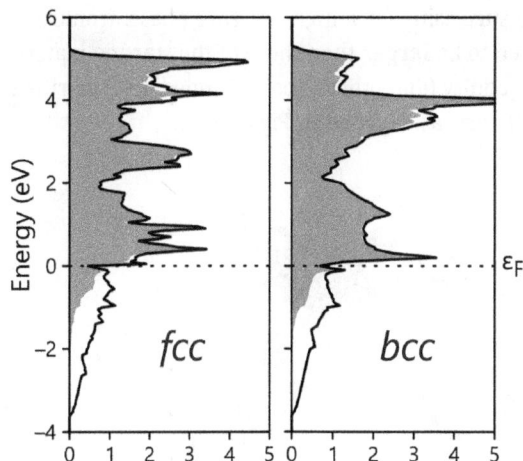

Figure 4.52: Comparison of the electronic DOS of *fcc*-Sr (left) with *bcc*-Sr (right) based on a plane-wave PBE calculation with the 4d participation shaded in gray.

of Sr is 2.15 Å in the *fcc* and 2.10 Å in the *bcc* structure. While one is tempted to assume another change in electronic **configuration** to be responsible for that, the change could not be more subtle. As the gross orbital populations projected from plane-wave DFT by LOBSTER evidence, the configuration is $4s^{1.97} 4p^{6.00} 5s^{1.27} 4d^{0.77}$ in the *fcc* structure, and this changes to $4s^{1.97} 4p^{6.00} 5s^{1.28} 4d^{0.76}$ in the *bcc* structure, so there is a minute (on the order of 1%) redistribution in the 5s 4d regime only, but it is clearly visible from the shaded 4d participation in Figure 4.52, and this leads to a higher DOS at the Fermi level for *bcc* (0.697) than for *fcc* (0.456), so **more metallicity,** in other words. At the same time, the increase in bond order may be quantified from ICOBI values. In the *fcc* structure, we have 12 nearest neighbors, and the ICOBI sum is 1.46, whereas in the *bcc* structure, there are 8 + 6 nearest and second-nearest neighbors, and here the ICOBI sum is 1.54, a 5% increase in bond order, also **more covalency.**

The "activation" of a higher oxidation state is not uncommon in high-pressure research, in particular for the experimentalist. For example, phases such as Cs_2CuF_6 containing Cu^{4+} demand high fluorine pressure (Müller, 1987); otherwise, everything stops at Cu^{2+} or possibly Cu^{3+}. Likewise, gold chemistry is usually limited to Au^{3+} but compounds such as $CsAuF_6$ (Müller, 1987) and AuF_5 (Hwang & Seppelt, 2001) obtainable via high fluorine pressure contain Au^{5+}. This is still in the MPa range but synthetically (fluorine!) quite an achievement. And yet, the GPa pressure range comes with new surprises.

The year 2004 saw a fascinating discovery of a new type of platinum "nitride" dubbed "PtN" obtained by a direct reaction between elemental platinum and molecular nitrogen, N_2, at a pressure of 45–50 GPa and activated by laser heating, the temperature being more than 2,000 K. The structure was determined to be the zinc blende type with fourfold coordination of Pt and N, and an extraordinarily high bulk modulus of 372 ± 5 GPa which would not be surprising for a high-pressure phase. Ironically, however, PtN's molar volume appeared to be **larger** than those of the starting materials, Pt and $\frac{1}{2}N_2$, a little paradox. Such oddity (and others, too) is immediately clarified by the results of electronic-structure theory, as depicted in Figure 4.53.

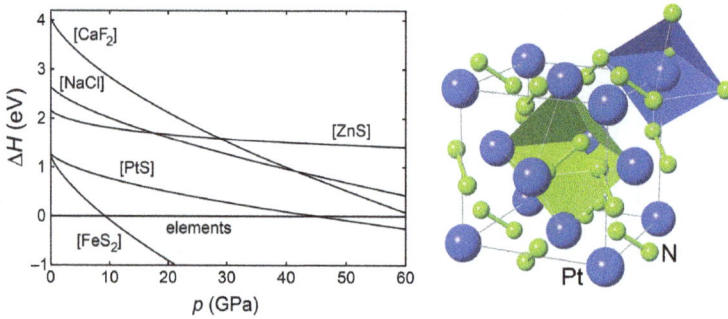

Figure 4.53: Course of the formation enthalpy of FeN and FeN_2 as a function of the structure type and pressure based on PAW plane-wave DFT (left) and crystal structure of PtN_2 (right).

Quite obviously, a phase dubbed "PtN" will never become stable adopting the [ZnS] structure type, and even the [PtS] type is strongly disfavored compared to the 1:2 composition, that is PtN_2. Hence, "PtN" could hardly exist, both composition and structure were likely to be incorrect (von Appen et al., 2006). Alternatively, the pyrite [FeS₂] structure type is favored, and subsequently the correct composition (PtN_2) and the correct structure (pyrite) were found by an independent study (Crowhurst et al., 2006) although the naming did not please the chemists since PtN_2 is **not** a nitride.

Because PtN_2 contains N–N dumbbells, understanding the structure, the composition, and also the high-pressure behavior is intimately connected with understanding the chemical bonding in this unit; there is no alternative, so we better start with the canonical molecular-orbital diagram of the N_2 molecule containing an $N{\equiv}N$ triple bond. It is shown in Figure 4.54, and can be compared with Figure 3.5.

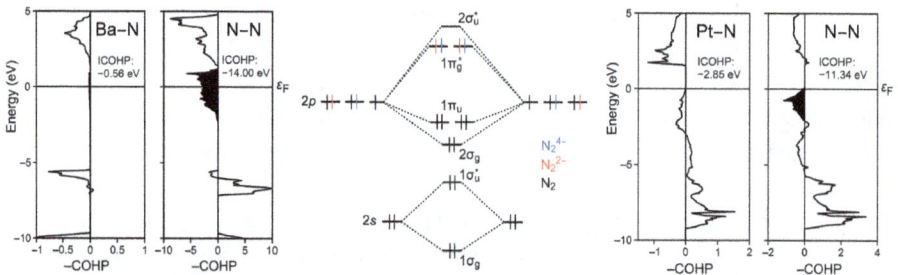

Figure 4.54: Iconic molecular-orbital diagram of N_2 with different electron fillings (middle) and metal–N as well as N–N chemical bonding in barium diazenide, BaN_2, (left) as well as in platinum pernitride, PtN_2 (right).

With an atomic valence electron count of 5, there are 10 electrons per N_2 unit, and they fill 5 molecular orbitals, the bonding/antibonding $1\sigma_g$ and $1\sigma_u^*$ (which cancel each other) and the bonding $2\sigma_g$ plus the two $1\pi_u$ (we saw them before in Figure 3.5); hence, there is a triple bond with a very short 1.10 Å distance and an enormous dissociation enthalpy of 946 kJ mol^{-1}. For two additional electrons as in N_2^{2-}, the degenerate antibonding $1\pi_g^*$ molecular orbitals get filled, hence, mimicking the electron configuration of the O_2 molecule, corresponding to an N=N double bond with 1.23 Å. This situation is found in barium **diazenide**, BaN_2, also made by high-pressure synthesis. If we were to add two more electrons and completely fill the $1\pi_g^*$ as in N_2^{4-}, the electron configuration would then correspond to F_2 or an N–N single bond with 1.41 Å, as found in the aforementioned platinum **pernitride**, PtN_2.

And this is also what the electronic structures of BaN_2 and PtN_2 (see again Figure 4.54, left and right) show. In the first case, we have moderate Ba–N bonding and half-filled antibonding levels for N–N, so this species is $Ba^{2+}N_2^{2-}$, simply put. In the second case, we have stronger Pt–N bonding but fully antibonding N–N interactions, so the phase corresponds to an electronic situation of $Pt^{4+}N_2^{4-}$. And **this** explains why the calculated bulk modulus of BaN_2 ($B_0 = 46$ GPa) is large due to the high molecular stiffness of N_2^{2-}: any outer pressure directly competes with the antibonding $1\pi_g^*$ being half filled, so the experimentalist presses against N–N antibonding molecular orbitals. And in PtN_2 (256 GPa) it is even worse because the $1\pi_g^*$ is completely filled (Wessel & Dronskowski, 2010).

Hence, the high-pressure chemistry and also physics of those diazenides (N_2 charged −2) and pernitrides (N_2 charged −4) are entirely dominated by the **complex** nitrogen anion (compare with Section 4.3). And it makes chemists tremble when BaN_2 or PtN_2 are dubbed "nitrides" because the name does not fit the electronic structure. A nitride implies an N^{3-} anion but "$Ba^{6+}(N^{3-})_2$" or "$Pt^{6+}(N^{3-})_2$" are grossly incorrect, in contrast to $Ba^{2+}N_2^{2-}$ and $Pt^{4+}N_2^{4-}$ which hit the nail on the head.[35]

Given proper understanding, it is rather straightforward to predict other species such as LaN_2 with N–N = 1.30 Å whose electron count is best described as $La^{3+}N_2^{2-}e^-$, a metallic material, electronically similar to the phase LaC_2 crystallizing in the [CaC_2] type (Wessel & Dronskowski, 2010). Shortly after, it was made by shock-driven decomposition of $LaNO_3$ (Tschauner et al., 2013), yielding N–N = 1.31 Å. Also, the transition-metal phase FeN_2 was predicted (compare with [NiAs]-type FeN before) (Wessel & Dronskowski, 2011) and also made in a heated diamond-anvil cell (Laniel et al., 2018), N–N = 1.31 Å. The charge assignment of "$Fe^{3+}N_2^{3-}$" is not in full harmony with the Fe–N distances; instead, $Fe^{2+}N_2^{2-}$ would possibly be more appropriate.

Such interpretation agrees with the preceding finding of the ternary phase Li_2Ca_3-$(N_2)_3$, containing only alkali and alkaline-earth metals. This compound also stems

---

35 Those who really believe that BaN_2 or PtN_2 should be called "nitrides" should also call H_2O_2 = H–O–O–H simply "water" instead of hydrogen peroxide. Water is H_2O, though.

from a high-pressure high-temperature (9 GPa, 1,023 K) synthesis of an azide mixture of LiN_3 and $Ca(N_3)_2$ and turns out as being a metallic diazenide (Schneider et al., 2013). The electronic structure of the complex anions (there are several ones in the structure) is given in the center of Figure 4.55, at high and normal pressures, to be compared with the iconic MO schemes on the left and right.

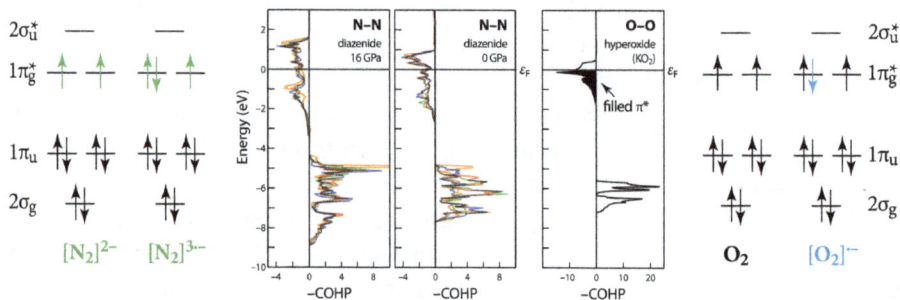

Figure 4.55: Schematic MO-based N–N (left) and O–O (right) bonding as a function of anionic charge together with COHP chemical-bonding analysis (center) of $Li_2Ca_3(N_2)_3$ at different pressures and of KO_2.

A rather naive consideration of the refined N–N distances between 1.34 and 1.35 Å might suggest the presence of N_2^{3-} but this is not the case, as evidenced by ESR, magnetic, and also electric conductivity measurements. First, unlike in the hyperoxide $KO_2 = K^+O_2^-$, the electron filling of the $1\pi_g^*$ levels does not comply with 75% but 50% occupation and, second, there is also no radical electron by ESR. Third, the vibrations of the anions agree with the N_2^{2-} formulation, so the two surplus electrons are delocalized within levels contributing to **metallic** character (Schneider et al., 2013), and we may describe the high-pressure phase as $(Li^+)_2(Ca^{2+})_3(N_2^{2-})_3(e^-)_2$ using the "ionic limit."

To put things into perspective, complex anionic dimers similar to N_2^{2-} have been well-known in solid-state chemistry for eternities, and they may be made **without** external pressure. For example, the simple calcium "carbide" phase CaC_2 is an industrial high-temperature product, and the ionic crystal-chemical formulation reads $Ca^{2+}C_2^{2-}$ so it really must be a calcium **acetylide** with a C_2^{2-} electron count just like in neutral N_2; hence, there should be a C≡C triple bond, and there is. Also, there are metal-rich phases such as $Y_2Br_2C_2$ with a Zintl-like electron balance of $(Y^{3+})_2(Br^-)_2C_2^{4-}$, so one expects a $C=C^{4-}$ anion resembling $N=N^{2-}$, correctly so, with fascinating physical consequences in terms of chemically controllable superconductivity (Simon, 1997), and some of its details are covered in Appendix I.

As high-pressure phases are usually associated with more condensation (as in the case of SiO_2 phases or ice X), we might have another look at the chemistry of complex nitrogen-based anions, for example, the carbodiimide anion (see Section 4.3). For illustration, let us focus on Figure 4.56, showing a number of simple and complex anions of nitrogen.

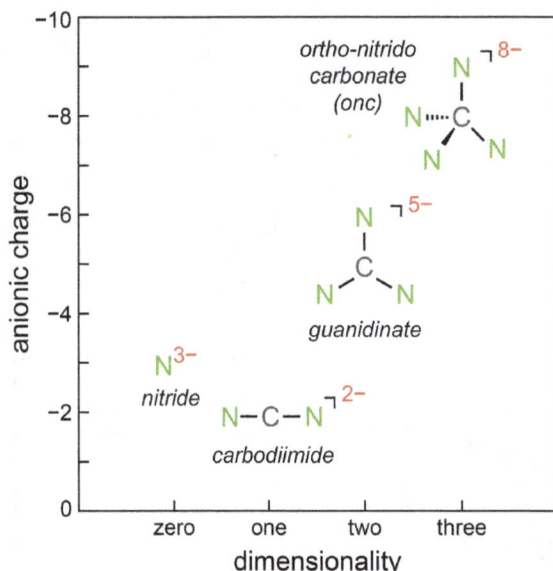

Figure 4.56: Shape of (complex) nitrogen-containing anions as a function of charge and dimensionality.

The nitride anion, N^{3-}, is zero-dimensional but the carbodiimide anion, NCN^{2-}, is elongated, hence one-dimensional. The fully deprotonated guanidinate anion, CN_3^{5-} is two-dimensional (triangular), and the ortho-nitrido carbonate anion, CN_4^{8-}, should be three-dimensional (tetrahedral). And it does not take a lot of fantasy to imagine that, under compression, $N^{3-} + NCN^{2-}$ should yield CN_3^{5-} while $N^{3-} + CN_3^{5-}$ should yield CN_4^{8-}. And in our times, all that can be computationally explored, as depicted in Figure 4.57 (top). Evolutionary algorithms predict species such as $TaCN_3$ (= $Ta^{5+}CN_3^{5-}$) and Hf_2CN_4 (= $(Hf^{4+})_2CN_4^{8-}$ whose C–N and also metal–N chemical bonding increases in strength (measured by ICOHP) as the pressure and, hence, condensation, increases.

Likewise, pressure polymorphs of a phase dubbed Hf_2CN_4 should exist, at least they are easy to detect from energy–volume diagrams. Expressed as enthalpy–pressure diagrams (see Figure 4.57 right), a nitride carbodiimide containing both N^{3-} (2×) and NCN^{2-} should exist as Hf_2CN_4 at standard pressure. At higher pressure, the more condensed nitride guanidinate polymorph containing N^{3-} and CN_3^{5-} should prevail, with the same Hf_2CN_4 composition. At highest pressures, the fully condensed ortho-nitrido carbonate form with only CN_4^{8-} will exist, all thanks to the ever-increasing covalent bonding (Luo et al., 2021). Let us now hope for experimental confirmation.

And something similar is also predicted for the simple $BeCN_2$ phase, still unconfirmed, a very interesting subject in terms of topology and stability by pressure (and also temperature), quickly summarized in Figure 4.58.

The two energetically lowest polymorphs are two-dimensional, almost graphite-like, and there is a high-temperature carbodiimide and one high-pressure wurtzite

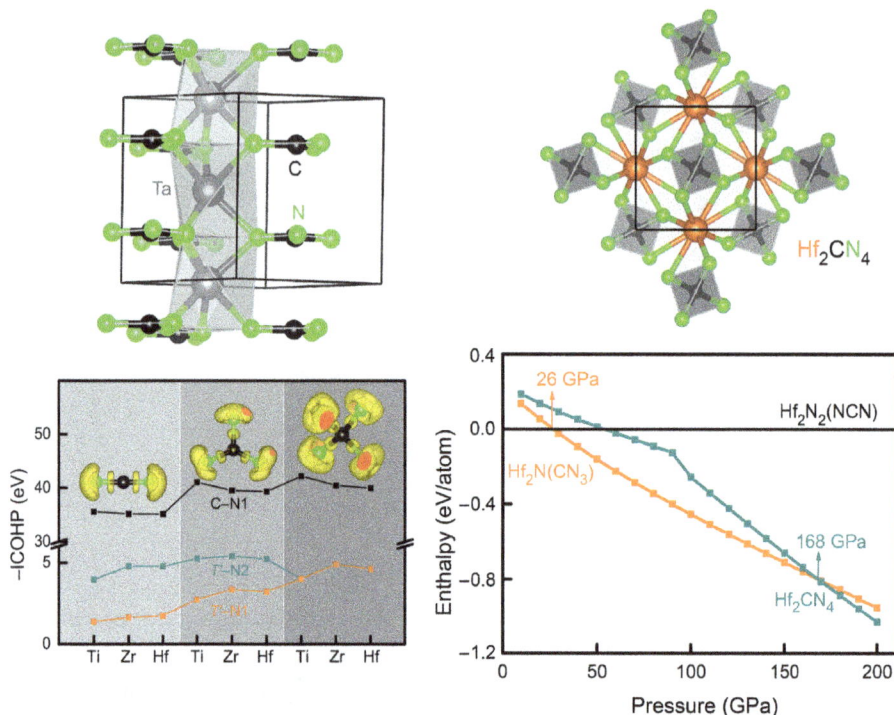

Figure 4.57: The predicted crystal structures of $TaCN_3$ and Hf_2CN_4 (top) as well as their C–N and metal–N bonding in terms of ICOHP and ELF as a function of pressure (bottom left) and enthalpy–pressure diagrams (bottom right) of the pressure polymorphs of Hf_2CN_4.

phase; all their stabilities can be rationalized, at least to a certain degree, by the amount of covalent C–N and, to a lesser degree, also Be–N chemical bonding (Luo et al., 2022).

And let us now return to the general assumption from the very beginning saying that pressure effectively metallizes everything, including hydrogen, by increased condensation.[36] This reminds us of the question asked to freshman chemists what will happen if a close-packed metal *M* in the *fcc* structure (with optimum space filling of 74%) gets compressed. Well, the volume will shrink but the structure will not change because *fcc* is already close-packed, and there cannot be a closer packing. That sounds right but it is wrong. Consider the simple metal Na crystallizing in the non-close-packed *bcc* structure type. Given sufficient pressure of about 65 GPa, it will turn into the *fcc* structure, and beyond that other funny things happen which we will ignore for the moment. At about 200 GPa, however, Na becomes **insulating** and **transpar-**

---

36 Another example is elemental krypton for which a computational study predicts metallicity beyond 500 GPa (Grochala et al., 2007).

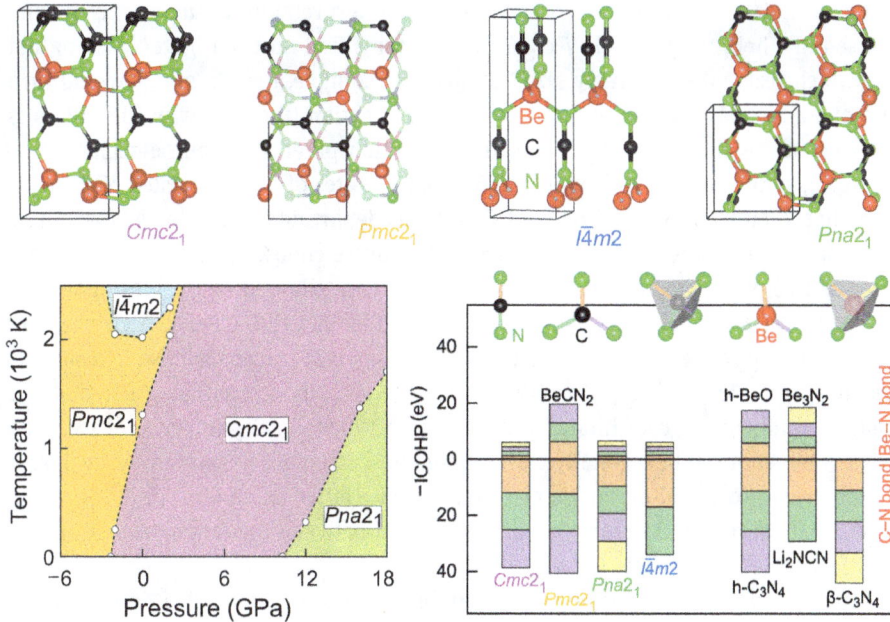

Figure 4.58: The predicted crystal structures of the four energetically lowest $BeCN_2$ polymorphs (top) and their temperature–pressure stability ranges as well as their C–N and Be–N bonding (color codes refer to the colored polyhedral sketches above) as compared to BeO and Be_3N_2 (bottom).

ent, the consequence of a massive structural reorganization adopting a [NiAs]-related atomic arrangement, depicted in Figure 4.59.

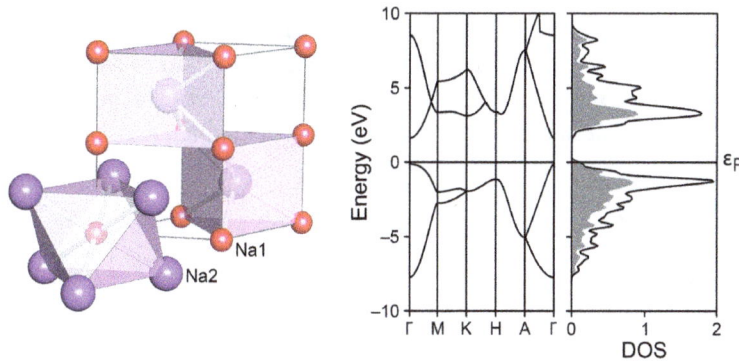

Figure 4.59: The crystal structure of transparent sodium (left) and its band structure and DOS (right), the shaded DOS part belonging to "anionic" Na, and the open part to "cationic" Na.

There is a bandgap, interpreted as a result of strong $3d–3p$ mixing, and even involvement of $2p$ core orbitals; at some point in the crystal structure, charges accumulate

(Ma et al., 2009). An alternative, possibly simpler interpretation would be this: if compressed like hell, the formerly *fcc*-packed Na atoms find an even better packing but one which rests on **differently sized** atoms. For example, two Na atoms would turn into one larger $Na^{\delta-}$ anion and one smaller $Na^{\delta+}$ cation (the energy used for that is the sum of the electron affinity and the ionization potential if one electron is exchanged), so the smaller $Na^{\delta+}$ would go into empty sites of a close-packed $Na^{\delta-}$ structure, that is, either the [NaCl] or the [NiAs] types departing from *fcc* or *hcp* would be the final result. In the case of transparent Na, nature chooses [NiAs] (a so-called reconstructive phase transition), and the Na atoms are charged ±0.20; admittedly, the charge transfer is much smaller than ±1 but the idea is valid. Clearly, high pressure leads to the **opposite** of metallization. That is to say that transparent Na is a **salt** between two Na ions of slightly different charge, grossly simplified, such that the smaller, octahedrally coordinated Na cation may "hide" in the empty sites (with no extra volume) of the hexagonally close-packed host structure generated by the larger Na anion with trigonal-prismatic coordination, the latter using up a little more volume. So, the structure mirrors the size difference between the larger trigonal prism and the smaller octahedron.

Indeed, such aforementioned qualitative analysis of atomic size depending on pressure can be carried out quantitatively given the right approach dubbed DFT chemical-pressure analysis, focusing on the impact of the atomic space requirements. One would intuitively assume that the spatial optimization of covalent bonding will compete with steric repulsion (i.e., Pauli repulsion between filled, core-like atomic orbitals), so **local** pressures in solids arise, and nature finds a compromise as given by real but often nonideal interatomic distances (Fredrickson, 2012).

The story of the synthesis of Na_2He, the first true compound of helium, resembles the transparent sodium case. At a pressure above 113 GPa, a mixture of helium and sodium inside a diamond-anvil cell gives rise to an X-ray diffraction pattern that corresponds to the anti-fluorite structure type, as shown in Figure 4.60. Apparently, there is an exothermic compound in which He **binds** to Na, at least at first sight, despite the $1s^2$ noble-gas configuration for helium. What is really going on in that phase?

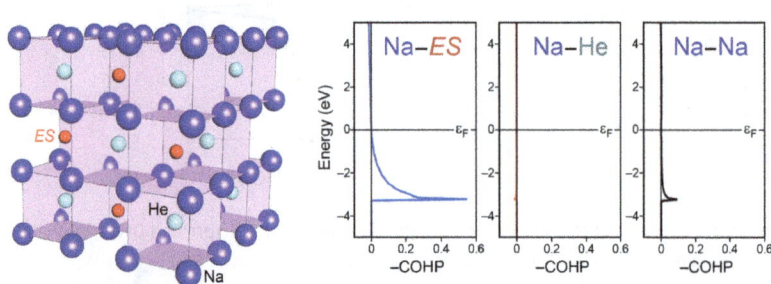

Figure 4.60: The crystal structure of Na_2He (left) and the corresponding COHP chemical-bonding analysis (right). The red "empty site" occupied by two electrons is designated as *ES*.

In the crystal structure, there are both Na cubes in which He atoms are found, and there are also empty Na cubes, in a 1:1 ratio. The material is insulating, salt-like, so to speak. A moment reflection yields (remember the story of transparent Na) that the pressure may be so high that the only valence electron belonging to Na, the $3s^1$ electron, can eventually be squeezed, mechanically, out of Na to generate Na^+ whose ionic radius for sixfold coordination (1.02 Å) is much smaller than the metallic radius (1.8 Å), so a lot of space is saved and the pressurized material may shrink in volume. And then a pair of two electrons (from $2\,Na \rightarrow 2\,Na^+ + 2e^-$) can move into the empty cubes made up of eight Na^+ cations, which is the rather simple idea. Fortunately, this is exactly what is found by electronic-structure theory, now carried out using local orbitals (LMTO theory, it allows for numerical orbitals = "partial waves" at an empty atomic site), depicted in Figure 4.60 (right). The COHP shows that there is practically no Na–Na interaction because the valence electron has already left Na at 113 GPa. Also, there is no Na–He interaction because both species, He and Na^+, have noble-gas configurations, [He] and [Ne]. However, there are strong interactions between the Na^+ ions at the corner and the two paired electrons in the center of the cube, designated as *ES* (= empty site), and this yields a lot of stability.

Electrostatically expressed, there is a Madelung energy term generated between a cation (Na^+) and an anion which only consists of two electrons ($2e^-$). That being said, we are witnessing a so-called **electride** phase because the electrons no longer reside where they originate from or belong to (Dong et al., 2017). So, the He atom simply acts as a "spacer" to keep the charges away from each other, and by doing it indirectly, it somehow "binds" to Na. Weird, in a sense, we are not used to think like that.

Finally, let us come back to a regular and ubiquitous chemical compound which can only be seen (in terms of a crystal structure) at high pressures. The unstable carbonic acid, H_2CO_3, is the simple product of a reaction between water and carbon dioxide, and we witness its decay product in everyday soda water, either from the bubbling CO_2 or the slightly acidic taste (H^+) due to H_2CO_3 being deprotonated, so HCO_3^- (hydrogen carbonate) is also in solution. Does H_2CO_3 **really** exist? There have been claims in the literature, based on clever syntheses and spectroscopic attempts, but as long as there is no crystal structure, only a few will believe that because **seeing** is believing; this is the reason why H_2CO_3 is often dubbed a "nonexisting" compound in the textbooks. And H_2CO_3 is important, for example, in the regulation of our blood's pH value. Close to 2 GPa, DFT shows that carbonic acid is enthalpically more stable than H_2O and CO_2. By using neutron diffraction and deuterium (D) instead of protium, it can be experimentally proven that pressure-stabilized D_2CO_3 becomes crystalline in the monoclinic system, as shown in Figure 4.61.

The D_2CO_3 molecules are arranged in dimers, and each one has two longer C–O bonds (1.30 Å) in the hydroxyl groups and a shorter C=O bond (1.26 Å) in the carbonyl group, simply because of the presence of ionic configurations (in valence-bond language) or due to π bonding with a bond order of ⅓ (in terms of MO theory), witnessed before in the doubly deprotonated guanidinate anion (see Section 4.2). This is also eas-

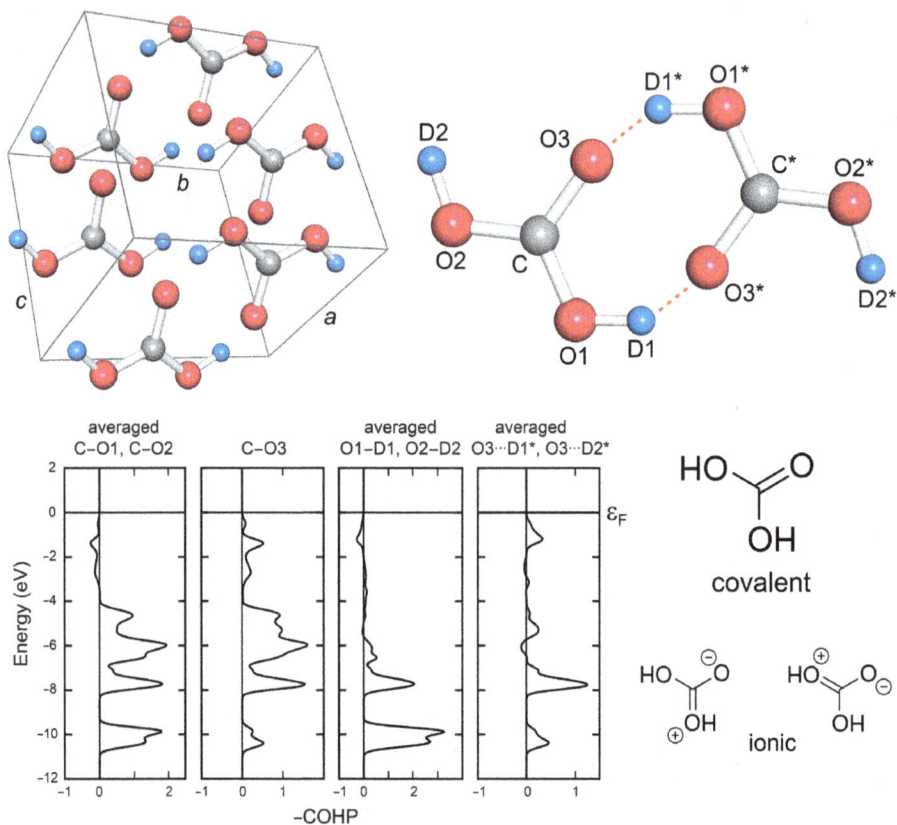

Figure 4.61: The crystal structure of deuterated carbonic acid and its molecular dimer (top) and the COHP chemical-bonding analysis of the various interactions in the crystal (bottom).

ily quantified from the COHP diagrams in Figure 4.61 (bottom). The O–D bonds are rather normal but they compete with extraordinarily strong O–D⋯O bridging bonds (see Section 4.4). A very puzzling finding is given by the wave function-derived Löwdin charges: C is charged +0.64, hydroxyl O is −0.42, carbonyl O is −0.50, and H is charged +0.35, and from this results an electrostatically defined Madelung energy of −980 kJ mol⁻¹, an enormously large amount comparable to rock salt. That does not mean, of course, that carbonic acid will evaporate as ions (likewise, MgO does not evaporate into Mg^{2+} and O^{2-} ions, no matter what is written in many textbooks) but it gives us an impression of the forces present in a molecular crystal with pretty short covalent bonds, given that there is a significant outer pressure (Benz et al., 2022).

Although we could continue with many more examples, it is time for some rest. A few points may be summarized as regards (high) pressure influencing chemical bonding in general:

- There are two well-known pressure rules, the first one relating to the behavior of one element aligning with its higher homologue and the second one highlighting the growing coordination number with pressure, the latter alluding to condensation behavior, well-illustrated by SiO_2 polymorphism and Pauling's third rule of crystal chemistry. These are good rules but not without exceptions.
- High pressure can cause electronic configurations to change, and it may also allow for higher oxidation states, not only for ionic materials. Quite to the contrary, even in presumably metallic materials, their behavior may at least be rationalized by different electronic configurations and, hence, different bondings.
- When complex anions such as pernitrides are part of solid-state materials, chances are pretty high that the entire pressure behavior is straightforwardly explainable from chemical-bonding arguments, in particular as regards bond distances, bulk moduli, and so forth, because pressure then acts on antibonding molecular levels.
- Likewise, pressure can determine the shape of complex anions and the degree of condensation, so pressure polymorphism goes far beyond what was known in the past.
- Opposed to the simple "pressure metallizes everything" traditional faith often found in solid-state textbooks, pressure can make things more ionic, otherwise highest-pressure phases such as transparent Na and Na_2He would be impossible to explain. This ionicity argument also relates to regular molecular compounds where pressure may induce unexpectedly large Madelung fields.

4.7 Multicenter bonding in solids

If one thinks of covalent bonding in elemental crystal structures composed of only main-group elements, the diamond structure type adopted by C, Si, and Ge immediately comes to mind, and that was the reason for introducing it in Section 4.1, nicely complying with the octet rule and filling all bonding levels, but with no electrons in antibonding ones. This situation mirrors stability, and it also serves as the prototype covalent bonding type with single atom–atom bonds and two electrons shared, the so-called two-center two-electron (2c–2e) bond. Going to the left side of the periodic table, one soon runs out of a sufficient number of electrons, hence electron socialism (= the metallic state, Section 4.5) sets in, simplifying things quite a bit. But what will happen if we allow for even more electrons and stay in the main groups? Take elemental tellurium, a late main-group VI or group 16 atom, the higher homologue of sulfur whose ground state is defined by S_8 molecules with S–S single bonds still allowing to fulfill the octet rule.

The crystal structure of tellurium, however, depicted in Figure 4.62, shows that Te, in its ground-state structure dubbed α-Te, contains infinite Te–Te-bonded helices, not simple Te_8 molecules, so there are two short and four longer Te–Te distances. Something similar is known for sulfur, because S_8 also polymerizes to infinite S–S-bonded helices at elevated temperatures. And starting from α-Te, one can imagine a less complicated simple-cubic (sc) Te structure, in which each Te is surrounded by six Te. In fact, a structure very similar to that is known at high pressure, in the GPa range (Section 4.6, pressure rules). Why does Te crystallize in the weird α-Te type under standard conditions but not in the sc-Te type?

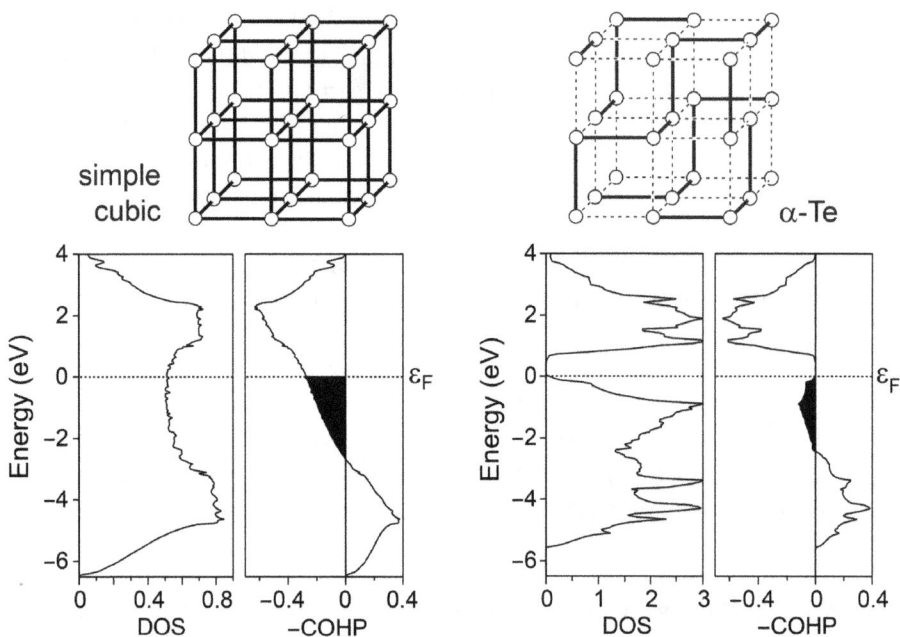

Figure 4.62: Sketches of tellurium's simple-cubic (left) and regular (right) crystal structure together with their DOS and Te–Te COHP analyses. The filled antibonding levels have been blackened.

A simple all-electron (LMTO) electronic-structure calculation even using the simplest functional (LDA) not only shows that the energetically higher lying sc-Te has become metallic with a finite DOS at the Fermi level but there are also strongly antibonding Te–Te interactions (in black) below the Fermi level (hence, occupied) against which we need to press using a few GPa; **this** is the reason for having to use high pressure. And upon releasing the pressure, the sc-Te type collapses into the α-Te type, replacing six equidistant 3.16 Å Te–Te bonds by two shorter (2.83 Å) and four longer (3.49 Å) ones, a **three-dimensional** Peierls distortion removing translational symmetry (Decker et al., 2002a). This rearrangement is driven by the annihilation of most formerly antibonding levels although **some antibonding levels still persist** below the Fermi energy (also in black), a consequence of the **too many** valence electrons available for Te. Something similar happens to the electron-poorer indium (Häussermann et al., 1999) as well.

To anticipate things a bit, we can already state that the physical consequences of these too many valence electrons (not only for Te, but also seen in a plethora of other phases) and their associated chemical bondings have made the fascinating research field of **phase-change materials** (PCMs) possible which allows for information storage and retrieval (say, in rewritable Blu-Ray disks) and other functionalities. To make a PCM work, our engineering friends operate with two solid-state phases of distinctly different physical properties (say, a metastable crystalline one and an amorphous one) to be used to encode one and zero bits, and rapidly switch back and forth between them, either encoding by laser (optical media) or current pulses (electronic media) and reading out given sufficient differences in reflectivity or conductivity (Deringer et al., 2015a). Even several decades after these often very complex materials had been used in everyday applications, additional research eventually showed that the simplest PCM is given by elemental tellurium itself, a fascinating discovery indeed (Shen et al., 2021), but easily understandable by looking at Figure 4.62.

Let us study the importance of the "right" electron count, not in an elemental material but in simple binaries, for example 2:3 compounds of other main-group or **almost** main-group elements, say, Sb_2Te_3 and Sc_2Te_3 (Rao et al., 2017). The Sc_2Te_3 composition is trivial to understand because Sc has a $4s^2\,3d^1$ electron configuration[37] and is a metal (which means low in EN), so it may and actually will provide all three electrons and become a Sc^{3+} cation, the ionic notion. Te is from main-group VI and will strive for Te^{2-}, so $(Sc^{3+})_2(Te^{2-})_3$ with [Kr] and [Xe] shells is a good starting point. And because of that Sb_2Te_3 will also correspond to $(Sb^{3+})_2(Te^{2-})_3$ but with one **decisive** difference: because Sb is from main-group V and has a $5s^2\,5p^3$ configuration with likewise five valence electrons, there will be two **surplus** electrons residing on Sb^{3+}, an electron **lone pair** as the chemists call it. Figure 4.63 exemplifies what that means in terms of electronic structure and bonding.

The electronic structure of Sc_2Te_3 (Figure 4.63, left) looks rather normal to us, simply because both atoms fulfill the octet rule, just like in diamond or GaN or any other material following this rule. That being said, there is a (tentative) bandgap between valence and conduction regions, and the entire valence band as seen through COHP and also two-center COBI is bonding, stabilizing the structure; the unoccupied antibonding levels are to be found in the conduction region. Although there is also a clear separation between valence and conduction regions, Sb_2Te_3, however, shows a characteristic fingerprint (Figure 4.63, right), **antibonding** in terms of Sb–Te, highlighted in red, and it is **occupied** because it lies below the Fermi level. In other words, the $5s^2$ "lone pair" located on Sb^{3+} and violating the octet rule is not very "alone" but it interacts antibondingly with Te, and **this** is what introduces trouble into the structure by

---

[37] In order not to mix up isolated with **chemically bonded** atoms, writing $(3d4s)^3$ for the electron configuration of Sc is another (probably better) option.

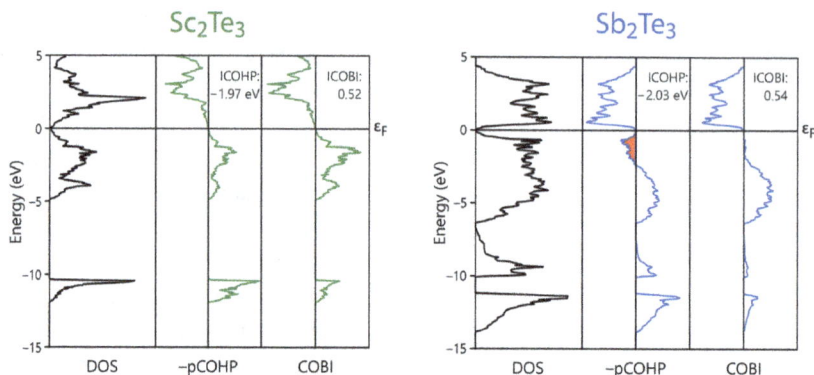

Figure 4.63: DOS, COHP and two-center COBI of Sc_2Te_3 (left) and Sb_2Te_3 (right) highlighting the difference in Sc–Te and Sb–Te bonding. Antibonding levels in Sb_2Te_3 have been emphasized in red.

some amount.[38] The "lone pair" character, on the other side, is visible from the fact that the two-center COBI is practically zero at that energy, the "lone pair" does not contribute to the bond order. Covalently, both Sc_2Te_3 and Sb_2Te_3 (COHP) are not very far from each other, and the two-center bond order is also similar (about 0.5, so a "half bond," see Appendix A) in both. One might even say that the electron-rich Sb_2Te_3, too, lets Sb bind to Te via two-center one-electron bonds, so-called singlet linkages (Sugden, 1930). For Sc_2Te_3, the Löwdin charges are $Sc^{+0.80}$ and $Te^{-0.53}$, and there results a Madelung energy of 801 kJ mol^{-1}, in addition to covalency. For Sb_2Te_3, however, the charges are insignificant, $Sb^{+0.18}$ and $Te^{-0.12}$, and so is the Madelung energy, only 39 kJ mol^{-1}. And yet, occupied antibonding levels are **only** found in Sb_2Te_3, the compound is too electron-rich, as said before.

Clearly, Sb_2Te_3 and Te resemble each other as regards their bonding properties close to the Fermi level. The similarities go further, however, directly seen from the ingredients of commercially available "GST" materials, an awkward abbreviation[39] of so-called semiconducting Ge-Sb-Te phases being used in rewritable optical media. Besides its true ground-state hexagonal crystal structure, a metastable "GST" phase with an idealized composition of "$Ge_2Sb_2Te_4$" exists, and it takes the simple [NaCl] structure in which the cationic (Na^+) sites are filled with Ge and Sb (2 + 2 = 4) and the anionic sites (Cl$^-$) by Te (4), so the composition fits the structure. And here is a most interest-

---

38 This "trouble" in terms of antibonding interaction is not too uncommon. For example, the compound In_4Br_7 (with In^+ and a $5s^2$ antibonding "lone pair," too) is so fragile that it decomposes under light and mechanical stress (Dronskowski, 1995).

39 When superconducting cuprates such as $YBa_2Cu_3O_7$ were discovered at the end of the 1980s, some communities decided to abbreviate them as "YBCO," acceptable in principle since each community uses its own language. There will be trouble, however, if one searches for oxidized (O) boron–carbon (B–C) networks bonded to yttrium (Y) in a structural database: suddenly thousands of falsely declared $YBa_2Cu_3O_7$ entries pop up. For superconductivity, see Appendix I.

ing observation: in the real, not in the idealized world, there are always plenty of Ge vacancies, ca. 20%, so "$Ge_{2-x}Sb_2Te_4$" would be a better composition because nature kicks out Ge atoms. Why? Kicking out Si from diamond-Si **costs** about 3.3 eV but here the materials stabilize themselves by Ge vacancies. Figure 4.64 based on electronic-structure chemical-bonding theory explains the cause.

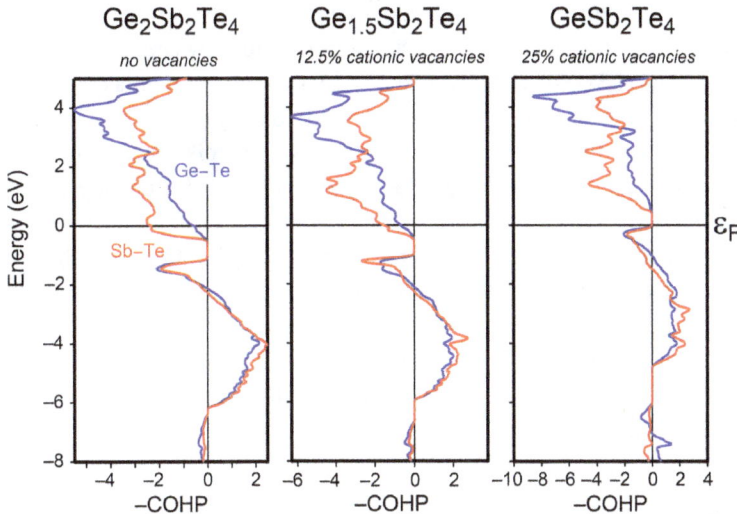

Figure 4.64: COHP chemical-bonding analysis of Ge–Te and Sb–Te interactions in various "GST" phases; the Fermi levels have been set to zero.

As witnessed from the Ge–Te and Sb–Te chemical-bonding interaction in $Ge_2Sb_2Te_4$ in Figure 4.64 (left), there are strongly antibonding levels below the Fermi level (so they are occupied), a consequence of the too high electron count. Among the three atoms, Ge is the least electronegative, so it is the most "cationic" atom that provides electrons to the most "anionic" one, Te; this notion is simplistic but it will serve well for the moment. Hence, if nature kicks out half a Ge atom, it **lowers** the electron count, and the Fermi level for $Ge_{1.5}Sb_2Te_4$ must also become lower, and fewer antibonding levels get filled, which is true (same figure, center). Taking out even more Ge atoms such as in $GeSb_2Te_4$ (right), there is even less antibonding. At the same time, with each Ge atom missing, one also loses Ge–Te bonding, so there must be a compromise between kicking out electrons (and annihilating antibonding levels) and leaving them in (and save some Ge–Te bonding). Unlike in the above simple computational model, in reality a phase close to $Ge_2Sb_2Te_5$ is found, also with Sb vacancies since Sb is also less electronegative than Te, it also serves as a "cation" (Wuttig et al., 2007). In fact, the well-known secret to making useful "GST" materials lies in the presence of elemental tellurium, an omnipresent ingredient, and the right number of **vacancies** needed to install the right electron count, see also below (Yamada, 2012). We already know that antibonding levels just below the Fermi level "drive" these materials.

And so the story goes. A too large electron count may lead to a "weird" crystal structure by Peierls distortion (Te case) or to a "weird" composition ("GST") to get rid of antibonding levels. And this can also be exploited by the restless synthetic chemist trying to **design** antibonding-free materials on purpose, by clever electron counting. Take InTe (= $In^{2+}Te^{2-}$) with one leftover electron on the formal In(II) atom and an electron configuration of $5s^1$. If we make this material **computationally** crystallize in the [NaCl] type (instead of its [TlSe] structure type, with one-dimensional motifs and tetrahedral coordination of In by Te, not shown), then a COHP diagram lets us easily spot the leftover electron on indium from the antibonding In–Te spike just below the Fermi level since structure and electron-rich electron count do not fit (not shown, do it yourself). InTe's unfortunate electron count of $3 + 6 = 9$ can be cured, however, by an alternative composition. A tripled stoichiometry such as In_3Te_3 does not change anything but if we were to substitute two Te atoms from main-group VI by two Sb atoms from main-group V, we would end up with the chemical formula of In_3Sb_2Te, and the electron calculus would read $3 \times 3 + 2 \times 5 + 6 = 25$, so the average electron count is 8.33, no longer 9, and the octet rule is only slightly violated. The chemical-bonding situation is shown in Figure 4.65.

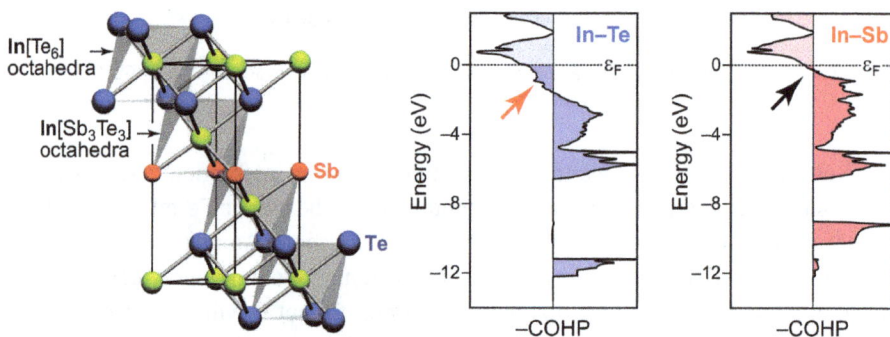

Figure 4.65: Crystal-structure model of $In_3Sb_2Te_2$ (left) in the [NaCl] motif and its In–Te and In–Sb chemical bonding (right).

Just like "GST," the phase In_3Sb_2Te usually called "IST" takes the rock salt motif but it is a metallic phase if crystalline, not a semiconducting one. Because of the lowered electron count – introduced on purpose – to optimize the composition, the In–Sb bonds are devoid of antibonding, and only a little In–Te bonding is left. In fact, antibonding is so small that, unlike "GST," there is no need to kick out other "cationic" elements. Hence, "IST" is devoid of substantial vacancies and shows better stability, and amorphous "IST" can stand distinctly higher temperatures than "GST" (Deringer et al., 2015b). The healing mechanism of "GST" and "IST" differs by the involved ions (cation vs. anion) and type (vacancy vs. substitution) but the underlying idea to annihilate antibonding levels is similar (Konze et al., 2019).

The prototype binary phase-change material is given by germanium telluride, GeTe, however, strongly violating the octet rule due to an electron count of $4 + 6 = 10$ relating to the neutral atoms. In the ionic limit assuming Ge^{2+} (and $4s^2$ leftover electrons) and Te^{2-} with the [Xe] shell, the calculus is also $2 + 8 = 10$, of course, no electrons get harmed. Structure and bonding are shown in Figure 4.66.

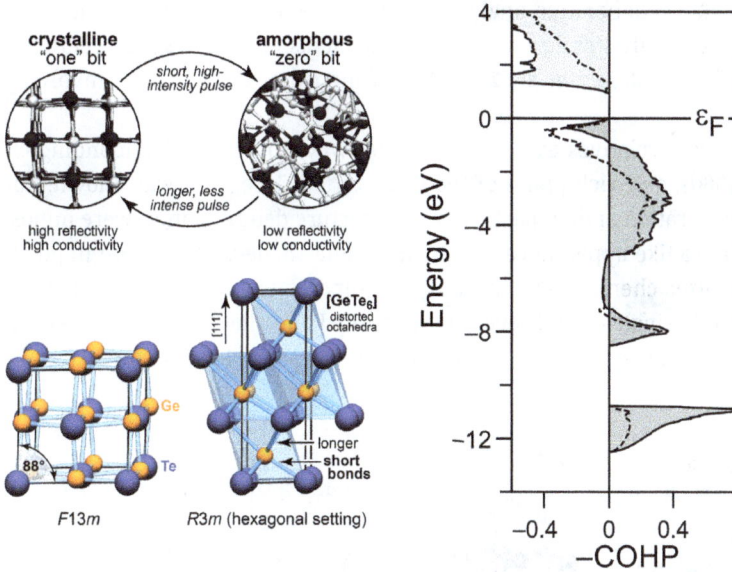

Figure 4.66: The principle of phase-change materials and the crystal structure of the binary prototype GeTe (left) together with the Ge–Te bonding just below the Fermi level (right) showing the course of the shorter (full lines) and longer (dashed lines) bonds.

Below the small sketch of the phase-change working principle (top left), there is the GeTe structure which comes **close** to the [NaCl] motif (bottom left). One usually shows the $R3m$ hexagonal setting of the rhombohedral phase but the $F\bar{1}3m$ pseudo-cubic setting is easier to comprehend, the angle of 88° indicating that the phase is **not** cubic (at higher T, it will become cubic again). In fact, there are not six **identical** Ge–Te bonds but three shorter ones, at 2.86 Å, and three longer ones, at 3.25 Å. The reader may have guessed that rhombohedral GeTe already underwent another three-dimensional Peierls distortion to lower the amount of antibonding, but there are still antibonding Ge–Te levels left (see Figure 4.66, right) (Deringer, 2014; Deringer et al., 2015a).

Now, the fascinating electronic structure of octet-rule-violating GeTe has important consequences. First of all, the phase is not perfectly 1:1 stoichiometric (also called **daltonide** by the chemists) but **berthollide** because even here one finds Ge defects (about $Ge_{0.9}Te$ from computation) which lower the Fermi level and heal antibonding, to some degree. Second, the mobility of Ge atoms in GeTe depends on the electronic structure and these aforementioned defects (Deringer et al., 2013; Maintz et al., 2016). Third, the surface structure of crystalline GeTe, in particular the α-GeTe(111) surface, is a function of the chemical bonding and the electron count (Deringer et al., 2012b). And finally the chemical bonding in amorphous GeTe does not differ much from the crystalline state, in principle, because the electron count rules. Nonetheless, despite octahedral motifs (we will come back to this later) the results of molecular-dynamics simulations plus chemical-bonding analysis highlight the peculiar role of homopolar Ge–Ge bonds which locally stabilize tetrahedral fragments but not the entire network (Deringer et al., 2014b; Maintz et al., 2016) by getting rid of Ge–Te bonds involving antibonding.

GeTe and its close relatives even provide other ways to heal the antibonding effects. Since the 1960s, a Se-rich phase of the composition $GeSe_{1-x}Te_x$ was known, but its phase-pure preparation and, in particular, its structure determination were impossible due to the mica-like appearance of the phase, and single-crystals were impossible to get. Then, after chemical-vapor transport (for this technique, see Binnewies et al., 2012) with GeI_4, single crystals eventually resulted and allowed to solve the crystal structure by X-ray diffraction and electron microscopy, depicted in Figure 4.67.

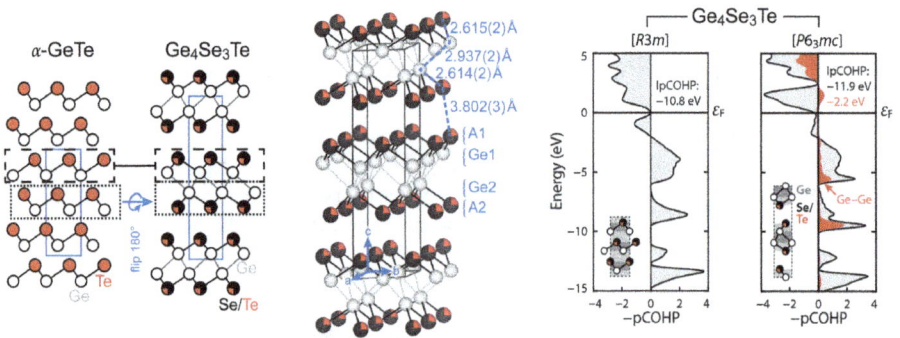

Figure 4.67: The crystal structure of GeTe and Ge_4Se_3Te (left), a perspective view into Ge_4Se_3Te (center), and Ge–Te and Ge–Ge chemical bonding in [NaCl]-like Ge_4Se_3Te and the real structure of the material (right).

Instead of adopting the Peierls-distorted almost rock salt GeTe structure, Ge_4Se_3Te (the correct composition) takes another one, a structure in which one GeSe/Te-layer has been flipped by 180° (left part of Figure 4.67). That is to say that Ge is no longer sixfold coordinated by Se (or Te) but by three Ge and three Se (or Te). The perspective view in the central part lets us understand that this leads to entire layers of Se (or Te) dubbed A1 and A2 packed on top of identical layers at 3.80 Å, so the mica-like behavior is immediately understandable. Ge_4Se_3Te is an almost two-dimensional material due to these mostly vdW interactions between the Se (and Te). DFT plus dispersion correction can even quantify them to be around 7 kJ mol^{-1} in strength. At the same time, unusually short Ge–Ge contacts at 2.94 Å result, and they are the clue to everything else (Küpers et al., 2017).

If Ge$_4$Se$_3$Te had to crystallize in the α-GeTe type (*R3m*), strongly antibonding Ge–Te and Ge–Se interactions would result, just like in α-GeTe, and that is unfavorable. Despite identical electronic configuration of Se^{2-} and Te^{2-}, the smaller size of Se lets one layer flip, and then the Ge–Ge distances can be sufficiently small such that they can take those electrons which would otherwise go into Ge–Te and Ge–Se bonds suffering from antibonding. In other words, electrons are **reshuffled** from antibonding levels into bonding ones, thereby generating new Ge–Ge partial bonds on the order of 34 kJ mol^{-1}. That isn't much but clearly better than having antibonding levels, so there is yet another way to get rid of antibonding, namely local structural rearrangement.

And this new behavior is mirrored not only from COHP, it is seen in the entire band-structure energy (sum of the Kohn–Sham eigenvalues), as presented in Figure 4.68, by using the density of energy (DOE) function.

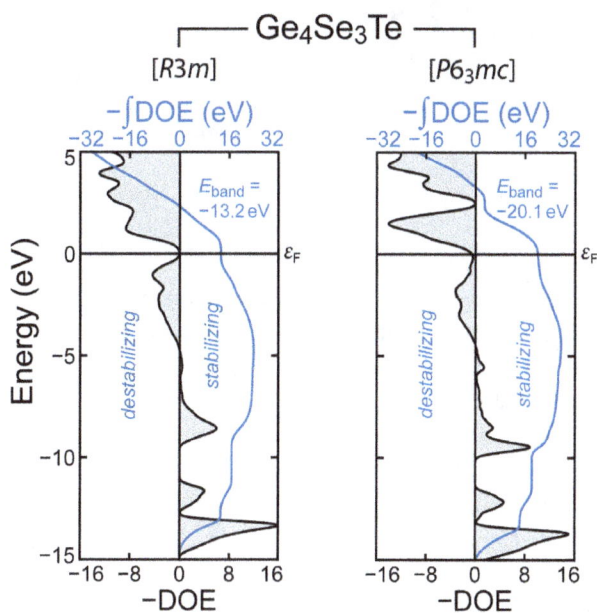

Figure 4.68: Density of energy (DOE) representation of the energetics in Ge$_4$Se$_3$Te adopting the [NaCl]-like (left) and the real structure of Ge$_4$Se$_3$Te (right).

The DOE indicates a rather huge increase in band-structure energy upon going from the α-GeTe type to the new layered Ge$_4$Se$_3$Te type (Küpers et al., 2017), not only by strengthened stabilizing interactions (both on-site, on the atom, and off-site, between the atoms) at low energies but also by weaker destabilizing interactions just below the Fermi level.

It is of crucial importance to stress, once again, the fitting or unfitting electron count. Often chemical-bonding ideas in this very realm of electron-rich materials violating the octet rule are based on arguments which relate to the physical properties such as transport behavior but whether or not a material is metallic, semiconducting, or insulating is **not** necessarily reflected in the chemical bonding, as exemplified by the comparison between SrTe (= $Sr^{2+}Te^{2-}$, i.e., $0 + 8 = 8$ valence electrons) and SnTe (= $Sn^{2+}Te^{2-}$, i.e., $2 + 8 = 10$ valence electrons), both crystallizing in the [NaCl] type. Their electronic structures and bondings (both DOS, projected COHP, and COBI) are presented in Figure 4.69.

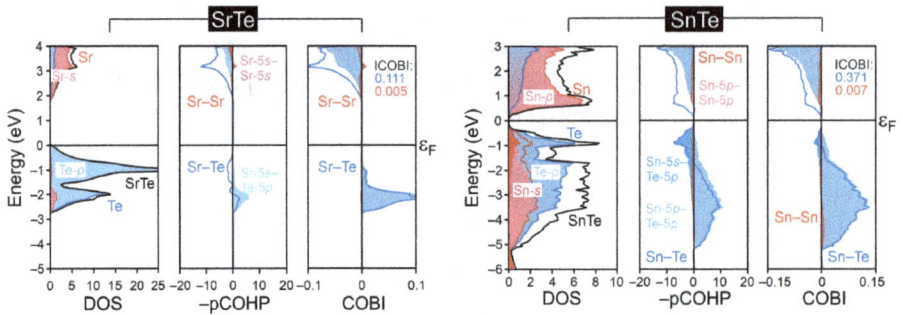

Figure 4.69: DOS, COHP, and COBI chemical-bonding analysis of various interactions in [NaCl]-like SrTe (left) and SnTe (right).

The II–VI-phase SrTe on the left fulfills the octet rule and is a regular semiconductor. The Löwdin charges of ±1.49 indicate a strong ionic character whereas the Sr–Te covalency as indicated by COHP is less important. Likewise, the Sr–Te bond order (ICOBI) is just 0.11, so little more than half a bond per Sr–Te contact, close to the value in CaO, another ionic prototype. The IV–VI-phase SnTe on the right, however, violates the octet rule and is **also** a semiconductor but with very different bonding. First, the Löwdin charges are close to zero (±0.08) which excludes significant ionicity. Quite to the contrary, COHP indicates strong covalency, and even ICOBI results in 0.37 or a total bond order of more than 2 (2.2) in a SnTe₆ octahedron, despite antibonding effects close to the Fermi level (Simons et al., 2021). So we may summarize and also emphasize that it is impossible to derive bonding from transport behavior, simply speaking. And even if Sr were replaced by Y, the one additional electron would make YTe metallic but the ionic chemical-bonding character would stay the same.

The rather unconventional bonding of Te or $Ge_2Sb_2Te_5$ or GeTe or SnTe (or PbTe) or In_3Sb_2Te or Ge_4Se_3Te which is seemingly mirrored from the phase-change behavior and several other properties such as extraordinarily large optical dielectric constant and a large Grüneisen parameter are worth mentioning. Even the appearance of the

[NaCl] type **despite** insignificant charge transfer (ionicity) seems rather strange. Why should GeTe, for example, not adopt fourfold coordination such as in the [ZnS] type? The octet-precise GaAs does it, why not GeTe violating the octet rule?

An unexpected observation can then be drawn from so-called projected force constants (pFC) as calculated for GeTe and other materials. Such pFCs are derived as follows: first one iterates toward self-consistency to arrive at a well-defined DFT ground state. The subsequent phonon calculations are carried out to yield the DFT-derived force-constant matrix and from that the phonon band structure and density-of-phonon states. And then one projects the force-constant matrix on the unit vector along a certain bonding direction, for example, along some first- or second-nearest interatomic distances; it might also be called bond-pFC (Deringer et al., 2015c).

If done for a "normal" covalent crystal such as diamond, the pFCs rapidly decay with interatomic distance: first nearest about 230 N m^{-1}, second nearest about 50 N m^{-1}, third nearest essentially zero. The situation is rather different for GeTe, GeSb$_2$Te$_4$, SnTe, and PbTe (violating the octet rule) on the one hand and CaTe and NaCl on the other, fulfilling the octet rule. A graphical overview is given in Figure 4.70.

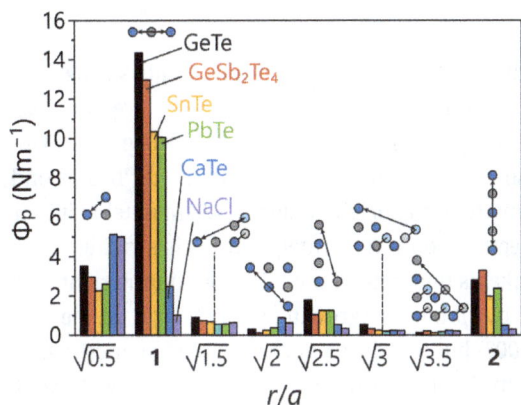

Figure 4.70: Projected force constants Φ_p as a function of normalized interatomic distances r/a in various phase-change and "normal" (= fulfilling the octet rule) materials.

The "octettians" CaTe and NaCl behave similarly to diamond, so their force constants (in blue and violet) also drop off rapidly as the interatomic distance increases, but not so in GeTe (black) and related materials: only for those compounds violating the octet rule one finds rather large pFCs for medium range, and even at very long distances; in particular, when atoms are **aligned**, the pFCs are extraordinarily strong. The unusual long-range force constants are present for all IV–VI compounds while neither CaTe nor NaCl show comparable effects. Something keeps the "non-octettians" together, and there is also a strongly directional influence in terms of **linearity** that alludes to σ interactions between p orbitals (Hempelmann et al., 2021).

A moment reflection of the VEC of GeTe (10, not 8) and the quantum chemistry of molecules allows for fascinating parallels. There is a fluoride of xenon, XeF_2, a linear molecule (Hoppe et al., 1962) also with a VEC of 10, and its molecular-orbital diagram (Coulson, 1964) is shown in Figure 4.71.

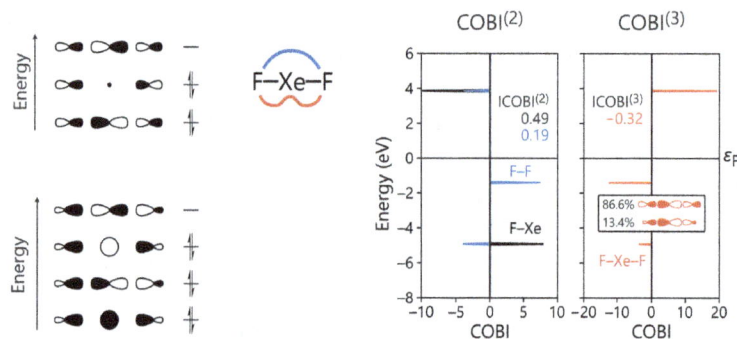

Figure 4.71: Molecular-orbital diagrams with/without Xe 5s orbitals in the XeF_2 molecule (left) and its two-center and three-center COBI analyses.

The MO diagram of XeF_2, first, without the Xe 5s orbital (top left) yields strong two-center Xe–F and also weaker second-nearest two-center F–F bonds; they are most easily quantified by the two-center $COBI^{(2)}$ plots (right) in which the integrated COBI yields bond orders of 0.49 and 0.19, so one "half" bond for Xe–F and one "fifth" bond for F–F (Müller et al., 2021), in harmony with the original model by Pimentel valid for the linear I_3^- anion with the same electron count (Pimentel, 1951). It is striking to see how the orbital **phases** on the left make us understand the COBI plot on the right. The model (bottom left) can be improved by sp mixing (growing with charge) on the central atom (Munzarová & Hoffmann, 2002) but this does not change the situation significantly, at least not for electrically neutral XeF_2. But there is also **three-center** F–Xe–F bonding, visible both in the MO diagram and in the three-center $COBI^{(3)}$ plot on the far right, and this yields an additional bond order of −0.32 (one "third" bond), the negative sign indicating "excess" valence electrons of a molecule violating the octet rule.

Are non-octettian phase-change materials such as GeTe held together by three-center or even multicenter bonding, this being the reason for their strange behavior, for example, in pFCs? A corresponding three-center $COBI^{(3)}$ analysis is displayed in Figure 4.72, based on plane-wave DFT calculations and the PBEsol functional plus D3 correction, but then projected onto local orbitals. To simplify things, we assume GeTe to be perfectly cubic, so regular octahedra around Ge and Te, and strictly linear Te–Ge–Te and Ge–Te–Ge units in all three directions. Hence, we only need to look at three-center interactions mediated by Ge (written as Te$\overset{Ge}{-}$Te) and those mediated by Te (written as Ge$\overset{Te}{-}$Ge).

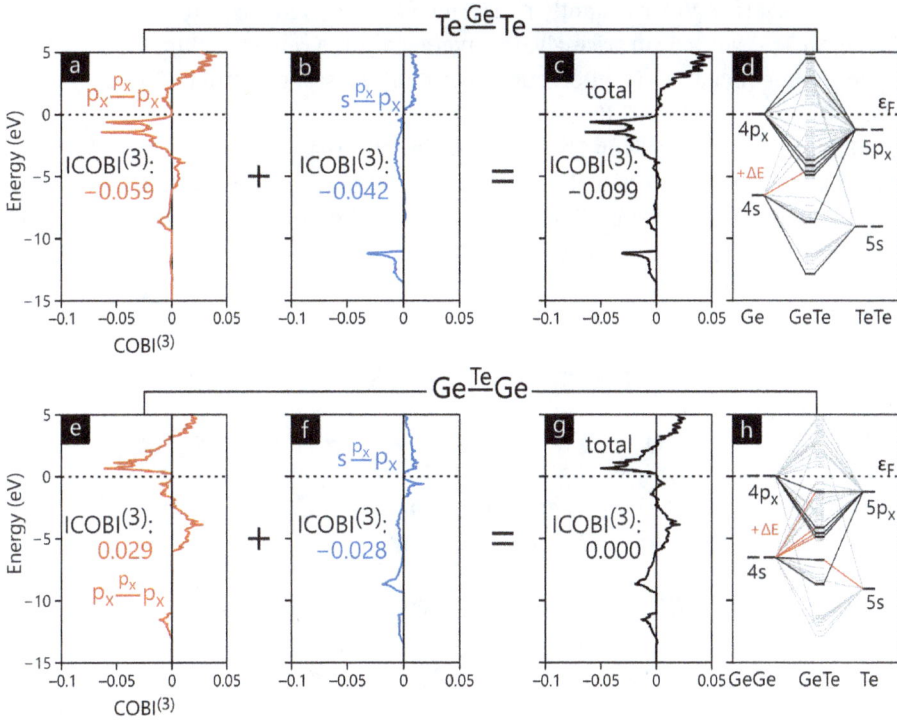

Figure 4.72: Three-center Te$\underline{\text{Ge}}$Te (mediated by Ge, top) and Ge$\underline{\text{Te}}$Ge (mediated by Te, bottom) interactions in GeTe and their analysis in "fragment" crystal-orbital-like representations.

Clearly, one finds 5*p*–5*p* and also 5*s*–5*p* interactions between the Te atoms, *mediated* by in-between Ge, and they add up to a three-center ICOBI of ca. −0.10 along one direction. Since there are three directions, the total ICOBI is about −0.30, just like in the XeF$_2$ case with the same electron count. Note that the mediating atom, Ge, locally violates the octet rule because of its 4*s*^{2} "lone pair," hence its role. As regards the 4*p*–4*p* and also 4*s*–4*p* interactions between the Ge atoms, mediated by Te, there are smaller effects but they fully cancel each other. Please also note that the mediating Te atom with an [Xe] configuration cannot mediate because it fulfills the octet rule (Hempelmann et al., 2022). And all that is also mirrored in the fragment crystal-orbital-like representation (Figure 4.72, right).

Such analysis is not restricted to GeTe but can also be carried out for the rock-salt-like phase antimony telluride, Sb$_2$Te$_3$, which consists of quintuple Sb–Te blocks separated by (due to its stoichiometry) vdW-like Te–Te gaps. Its electronic structure is easy to calculate (see beginning of chapter, we covered it already in Figure 4.63), and so is its phonon structure (Stoffel et al., 2015). And by chemically blending GeTe (with Sb$_2$Te$_3$), an entire solid-state mixture ranging from GeTe over Ge$_3$Sb$_2$Te$_6$ ("GST326") over Ge$_2$Sb$_2$Te$_5$ ("GST225") over GeSb$_2$Te$_4$ ("GST124") to Sb$_2$Te$_3$ can be synthesized, in which the average VEC constantly rises from $(4 + 6)/2 = 5$ for GeTe to $(2 \times 5 + 3 \times 6)/5 = 5.6$ for Sb_2Te_3. Not too surprisingly, the averaged three-center $ICOBI^{(3)}$ for the either Ge- or Sb-mediated Te–Te interactions constantly rises, too, from -0.10 over -0.12 over -0.13 over -0.14 to -0.16.

But let us concentrate on Sb_2Te_3: as alluded to already, there is a vdW-like structural gap between the terminal Te atoms, but the gap is **significantly** (12%) smaller than the sum of the vdW radii, despite no external pressure in the GPa range. And indeed, there is significant multicenter covalency beyond the gap, closing the gap, as displayed in Figure 4.73.

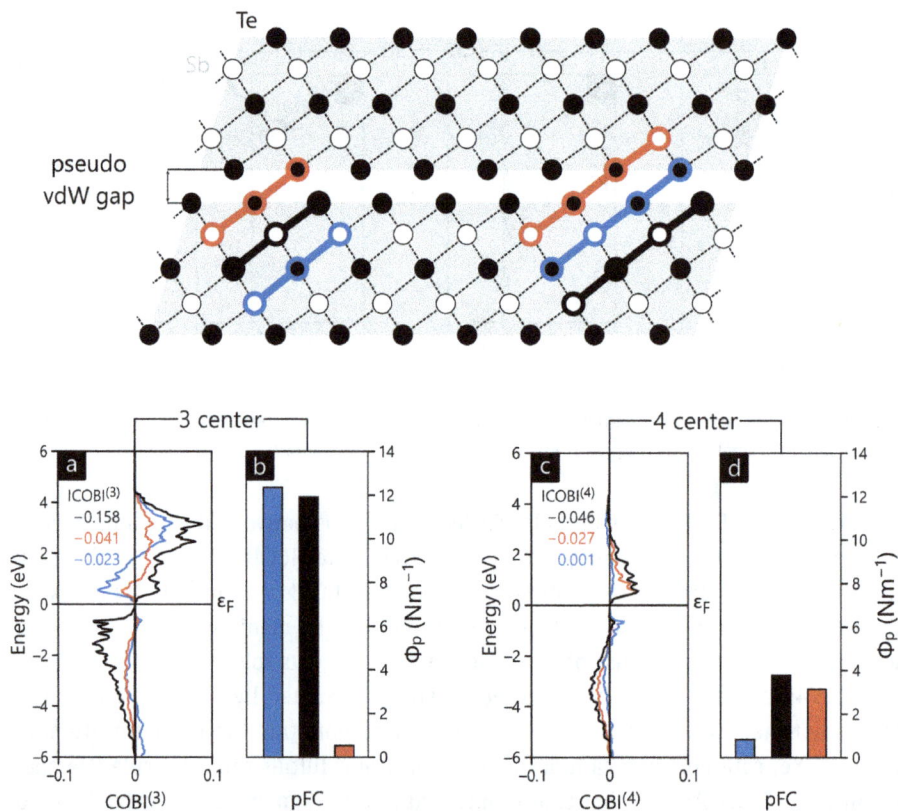

Figure 4.73: Three- and four-center bonding interactions in Sb_2Te_3 within the quintuple block and beyond the "van der Waals" gap (top) and the corresponding three- and four-center COBI and projected force-constant analysis (bottom).

As indicated in the upper part of the figure, three- and four-center interactions within the quintuple block but also beyond the gap exist, and their numerical $COBI^{(3)}$ and $COBI^{(4)}$ analyses as well as the pFCs are given below. Clearly, multicenter interactions also going under the name hyperbonding or hypervalent bonding (Lee & Elliott, 2021) exist, and they go **beyond** the gap, thereby explaining the fact that the vdW gap is too small, also seen for other octet-violating phases (Hempelmann et al., 2022). And there are also five-center interactions, nicely correlating with force constants. As such, multicenter bonding not only leads to an increased conductivity, long-ranged force constants, and atoms "sticking together" in atom-probe multiemission events, but also closes structural gaps between tellurium layers. In the case of the phase-change materials, a **too large** number of electrons is responsible, in clear contrast to delocalization for metals with **too few** electrons. Finding such effects took some time since local-orbital methods applicable to solids (i.e., COBI calculated via LOBSTER) had to be developed first.

Admittedly, there is nothing new under the Sun. The rather unusual bonding of the heavy nonmetals had been understood for the case of elemental Bi (Jones, 1934) in the early days of solid-state quantum mechanics already but such pioneering work is prone to be forgotten. Three-center bonding in (heavy) electron-rich complex anions has also been chemically analyzed (Landrum et al., 1997) almost three decades ago, and even extended solids have been carefully looked at (Papoian & Hoffmann, 2000), in particular as regards the element tellurium that is omnipresent in phase-change materials (Chivers & Laitinen, 2015). The exceptional case of tellurium is easily explained from its atomic properties (Jones et al., 2023), that is, the shape of the $5s$ and $5p$ orbitals and their mutual fit, in particular to the neighboring atoms.[40] Somewhat simplified, the chemical bonding in phase-change materials involves the simultaneous presence of two-center two-electron (2c–2e) and multicenter electron-rich (3c–4e) bonds, including weaker contributions over even more atoms.

The interested reader will have noticed that the circle to the chalcogen bonds from an earlier chapter is closed here, and that is no coincidence. The same effect shows up under a different name, in particular for a heavy chalcogen such as Te.

Multicenter bonding in solids is easy to summarize:

-   Unlike some elemental metals with **too few** electrons, there are also some elemental nonmetals with **too many** electrons, and the surplus electrons violating the octet rule show up antibondingly in COHP and may lead to Peierls distortion and/or structures that look more complicated than usual.
-   The same phenomenon is found for octet-violating compounds, and sometimes nature kicks out cations such as to lower the electron count; this can also be accomplished by a smart synthetic chemist through on-purpose rational stoichiometry.
-   In more complex structures, nature reshuffles electrons from antibonding into bonding levels, so there are cases of cation–cation bonding in order to locally optimize bonding.

---

40 For completeness one should add that Te becomes a superconductor at 4 GPa already, and further pressurization leads to the *bcc* type for Te, a superconducting phase with a transition temperature of 17 K whose band structure shows a characteristic "fingerprint" for superconductivity, regions of both electronic localization and delocalization; see Appendix I.

- Physical transport properties do not necessarily go hand-in-hand with the bonding character that may differ drastically despite, say, similar conductivity.
- Similar to multicenter bonding in electron-rich molecules, octet-violating elements and compounds engage in multicenter bonding such as to optimize bonding, easily visible by three-center, four-center, etc. COBI analysis. Such bonding also leads to regions of some weak covalency where only simple vdW bonding is expected.

5 What can be learned!

If this little book provides one important message, then it should be this: extracting and quantifying chemical-bonding information from first-principles electronic-structure calculations is a rather straightforward procedure, even when the original electronic-structure theory has been carried out using totally delocalized plane waves, therefore hiding the chemistry at the very beginning to achieve faster calculation. At a later stage, extracting all the "chemistry" from the quantum-mechanical result is **the** decisive step in understanding what keeps periodic materials together. Understanding solids (and molecules, too) from orbital interactions is simply a logical consequence of what has been done in the realm of molecules for decades already. Why shouldn't it be done for infinite molecules, that is, crystals?

It goes without saying that empirical tools to spot effects of chemical bonding (say, bond lengths, force constants, etc.) are used very often. When it comes to quantum theory, however, extracting or "projecting" the chemistry just requires a unitary transformation from the delocalized to the localized representation, easily and automatically performed using LOBSTER in case someone starts with plane waves. And the program is entirely free, it costs nothing except a little effort to download, install, and run the program. Admittedly, some chemical thinking is required, and this I have tried to teach you in this primer.

That being said, there is absolutely no need anymore to simply guess (often, incorrectly so) the bonding properties from a simple density of states (DOS) figure which does **not** contain orbital phases; note that the DOS is **blind** for such analysis. One can do much better, for example, a crystal orbital Hamilton population analysis besides the DOS, and then the interactions are directly visible, and so are the bonding proclivities hidden under the DOS. Remember that bonding is an **energetic** phenomenon. Likewise, there is absolutely no need to guess (often, incorrectly so) the strength of covalent bonding by whatever electron density or electron localization because the latter does **not** contain the bonding information; the density, no matter how accurate, and also the localization are simply lacking the phases. One can also do much better here, for example, by projecting the crystal orbital bond index along the interatomic vector, and this will yield not only the bond order but also, possibly, the information whether or not three-center interactions (or even more) are involved, impossible (I strongly believe so) to detect from the density.

In addition, the bonding tendency of electronic bands can be easily visualized such as to allow for, say, rational bandgap tuning for optoelectronic applications, there is no other quantum-mechanical way (except trial-and-error, but that belongs to the past). And let me assure you: there is an infinite amount of other information waiting to be discovered in today's and tomorrow's solid-state materials.

And now please do your own chemical-bonding analysis. Good luck! And if you enjoy it half as much as I enjoyed writing this book, a lot is gained.

https://doi.org/10.1515/9783111167213-005

6 Appendix: treats of general and quantum chemistry

When talking about chemical bonding in general, chemists profit from what they have learned (or should have learned) in standard university courses, and so some important principles well-known to those people are often **taken for granted** upon explanation; it is easier to explain what makes Beethoven's music great if you know what **classical** (as opposed to, say, **romantic**) music really is; the more you know, the easier the things will become. But many chemical concepts or, better still, "treats" are sometimes unknown to other highly intelligent people from related fields, simply because they (who may be physicists, engineers, computational scientists, etc., I respect them all) had other, more important things to learn at that time. And let us not forget that life is short, dear friends. In order to not force these people to read entire 1,000-pages textbooks of general and quantum-chemical content, here I briefly condense those chemical treats using very simple language and a few simplifications; the chemists will kindly allow that.

A The octet rule

The octet rule is an ingenious concept for counting the number of valence electrons in a plethora of molecules containing a few special light atoms all stemming from the first **long** period, also called second period (i.e., the elements between Li and Ne). In this very period, the elements C, N, O, and F strictly follow the octet rule. What is meant by that?

We first recall that all elements tend to achieve a noble-gas shell (at least we believe that), so H from the first **short** period (which only consists of H and He) forms the H_2 molecule with an H–H single bond (Chapter 3). Counting the shared two electrons for **each** H atom (yes, we are counting them twice, like a corrupt accountant but please do not worry), both the left and the right H atoms have two electrons, so both atoms correspond to the [He] configuration. It is shown in Figure A1. For the elements C, N, O, and F from the first **long** period, they tend to mimic the electron count of the Ne atom with the [Ne] configuration and **eight** electrons, so that is where the term **octet** comes from. Because of that, the tetrahedral methane molecule (CH_4) has a central C atom and four C–H single bonds "filled" by two electrons each such the electron count of C is $4 \times 2 = 8$, octet rule fulfilled. Likewise, in the ammonia molecule (NH_3), the N atom has one nonbonding electron pair and three N–H single bonds, so $2 + 3 \times 2 = 8$, octet rule also fulfilled. In the water molecule (H_2O), the oxygen atom carries two nonbonding electron pairs and forms two O–H single bonds, and this makes $2 \times 2 + 2 \times 2 = 8$. It works! And in the hydrogen fluoride (HF), molecule, there are three nonbonding electron pairs for F and only one F–H single bond, so we have $3 \times 2 + 1 \times 2 = 8$. In order to avoid

https://doi.org/10.1515/9783111167213-006

boredom, nature then invents double and triple bonds to still obey the octet rule whenever the number of atoms is too small. So, the O_2 molecule has an O=O double bond (four shared electrons) in addition to two nonbonding electron pairs for each O, so the electron count of either the left or right O atom is $2 \times 2 + 4 = 8$.[1] And in N_2 containing an N≡N triple bond (six electrons shared) and one nonbonding electron pair for each N, the calculus reads $1 \times 2 + 6 = 8$ for each N. And so the story continues, even for more complex molecules. Within the important disinfectant formaldehyde (H_2CO), a cute molecular triangle with a central C atom, a C=O double bond (with four shared electrons), and two C–H single bonds, the C electron count therefore reads $4 + 2 \times 2 = 8$. Try to sketch it yourself! Because almost the entire organic chemistry is made up of the elements C, H, N, and O, the octet rule is close to being a **law** for that discipline, and each ball-and-stick model of whatever organic molecule profits from that. Organic chemistry can't be too difficult (I am smiling).[2]

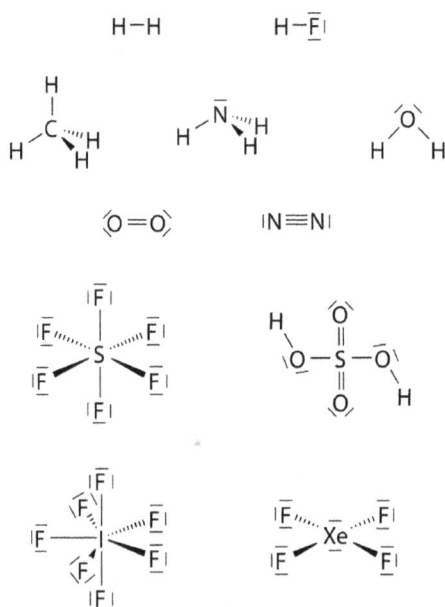

Figure A1: Lewis formulas of a few fundamental molecules. The professional chemists will note that "formal charges" have not been used, on purpose, because they will first show up in Appendix F. For triplet $^3O_2$ with two unpaired electrons, replacing one – by two · in the sketch would be kind to theory but is almost never done.

For the higher homologues of the elements of the first long period, say, those from the second or third long period, the octet rule does **not** hold, sulfur being a prominent

---

1 That is a **highly** simplified picture for a molecule involving a σ bond and a π bond, and the diradical character of triplet $^3O_2$ is rather weak due to electronic correlation (Borden et al., 2017).
2 The complexity, and difficulty, and elegance of organic chemistry is not a consequence of the chemical bonding but of the energetic proximity of a plethora of metastable molecules on the hypersurface spanned by C, H, N, O, and a few other atoms. Changing an organic molecule into another one is mostly a kinetic, not necessarily thermodynamic problem, mastered by the ingenious organic chemists.

example.[3] For instance, there is the perfectly normal sulfur hexafluoride (SF_6), molecule forming a perfect octahedron, and here the S atom is bonded to F via six S–F single bonds, so sulfur's electron count is $6 \times 2 = 12$, a massive conflict with the octet rule but the S atom simply doesn't care. And the same problem persists for sulfuric acid (H_2SO_4), at least when double bonds between S and O are formulated; the double bond is questionable, by the way, as covered in good textbooks of inorganic chemistry (Holleman et al., 2007; Greenwood & Earnshaw, 1997), and one may relieve the nonoctet pain by **formal charges** (see Appendix F). And there is a plethora of other violations from the octet rule whenever nature involves elements from higher periods, that is, the more "inorganic" the molecules become.[4]

The importance of the octet rule and **deviations** from the rule cannot be overstressed. In the first half of the twentieth century, fine synthetic chemists succeeded in making molecules such as BrF_5 (12 electrons for the central Br atom), IF_5 (also 12 electrons for the central I atom), and even IF_7 (14 electrons for the central I atom, imagine that). That is to say that the heavier the central atom, the stronger the tendency (or the tolerance) to deviate from the octet rule, possibly a geometric effect because there is simply more space around the central atom. The bonding partners should also be strongly electronegative. And **this** was one of the reasons for Bartlett, Hoppe, and contemporary chemists to search for fluorine compounds of the element Xe (as the immediate neighbor of the I atom) in the early 1960s (Bartlett, 1962). If iodine does not obey the octet rule, why should xenon, despite its electron octet in the neutral atom? And then the first binary compound of the noble gases, the magnificent XeF_2 (with 10 electrons for Xe) was eventually synthesized (Hoppe et al., 1962), followed by many others such as XeF_4 with 12 (Claassen et al., 1962) and XeF_6 with 14 electrons for xenon (Hoppe, 1964). The Royal Swedish Academy of Sciences should know why this wasn't worth a Nobel Prize in Chemistry.

B Electronegativity

According to the IUPAC definition, electronegativity is the tendency for an atom of a given chemical element to attract shared electrons (or electron density) when forming a chemical bond. The concept is usually assigned to Pauling who was the first to come up with a **quantitative** scale but the term itself was first used by Berzelius, and it is

---

3 Because the terms **short** and whatever **long** period may lead to confusion, an alternative designation has been proposed: **very short** period (H–He), **short** periods (Li–Ne, Na–Ar), **medium** periods (K–Kr, Rb–Xe), and **long** periods (Cs–Rn, Fr–Og).

4 It is almost forgotten these days that Sugden, more than 90 years ago, realized that the octet rule does not have to be violated in case "half" (two-center one-electron) bonds are involved, in addition to regular (two-center two-electron) bonds (Sugden, 1930). The concept of half bonds is perfectly valid, as we know today, but it was heavily attacked at that time and fell into oblivion.

clear that similar thoughts can be traced back to Avogadro already, amazing as it may seem (Jensen, 1996). Electronegativity is not an observable quantity, you guessed it already (see also Appendix D), but a fuzzy descriptor combining several chemical variables such as electrochemical potential, bond energy, and reactivities.[5]

Associating Pauling with electronegativity is due to the fact that, as alluded to before, he found a clever way to use the dissociation energies of the elements (thermochemical data, so to speak) and a tricky calculus to come up with a quantitative EN scale, so the method no longer was purely qualitative, which brought a lot of credibility in those misty times of the past. One needs a reference point, hydrogen came in handy, and this leads to the most typical EN scale with the electronegativities of H (2.20), Li (0.98), Be (1.57), B (2.04), C (2.55), N (3.04), O (3.44), and F (3.98). So, F is the most electronegative element with an EN of 4, easy to remember.

And then came other scales, for example, the one by Mulliken, in which he used the arithmetic mean of electron affinity and ionization energy, and this nicely works as long as the atomic data are known. The scheme by Allred and Rochow approximates the EN from the effective nuclear charge divided by the square of the covalent radius, and this also leads to a satisfying EN scale. And there are even more scales, and all of them work nicely and show the same qualitative trends; exceptions prove the rule, though. Not to forget, electronegativity can also be expressed by density-functional theory (DFT), as done by Parr and Pearson, from the first derivative of the energy against the electron number, and then it is termed **absolute** electronegativity, numerically equivalent with Mulliken's way to calculate EN (Pearson, 1988).[6] And the story continues; there is yet another, very recent electronegativity definition based on valence-electron ground-state energies (Rahm et al., 2019).

When thinking about the aforementioned $\partial E/\partial N$ derivative, it is clear that the same atoms composing different molecules or solids of the **same** composition but with **different** structures (isomers or polymorphs) should have different electronegativities, simply because the derivative will not only depend on the composition but also on structure; the E versus N diagrams should differ. Here is a simple example, just to illustrate that.

CaO crystallizes in the [NaCl] structure type, but there is also a more densely packed [CsCl] type, and one can imagine a more loosely packed [ZnS] type. Clearly, the [NaCl] ground state is lowest in energy and then followed by the [ZnS] type, eventually by the high-pressure [CsCl] type (Dronskowski, 2005). The tendency of Ca "to attract shared electrons or electron density when forming a chemical bond" to O **must** change, so the electronegativity of Ca also changes due to changing structure, that is obvious, and we can illustrate it from Löwdin charges and ICOBI values for the Ca–O bond, as extracted from plane-wave GGA calculations.

---

5 A theorist colleague calls chemistry the simplest natural science of **complexity**, and he is probably right.

6 In addition to that, an absolute **hardness** (second derivate of the energy against the electron number) can also be generated in the framework of DFT (Parr & Pearson, 1983).

In the [NaCl] ground state, Ca is charged +1.53, and the Ca–O ICOBI is 0.10, that is, a bond order of 0.60 for the CaO_6 octahedron, slightly more than half a bond. In the [ZnS] type, the Ca charge is a slightly larger +1.61 with an ICOBI of 0.14 and 0.56 for the bond order of the CaO_4 tetrahedron, slightly smaller than before. And in the [CsCl] type, the Ca is charged +1.59, with an ICOBI of 0.07 which corresponds to a 0.56 bond order for the CaO_8 cube, similar to the zinc blende type.

Clearly, CaO is **least** ionic and **most** covalent in the most stable [NaCl] type, a very interesting finding. And while Ca shares **more** electrons with O covalently but has attracted **fewer** electrons from O, the electronegativity of Ca must be unlike in the different polymorphs of CaO. In this respect, the IUPAC definition is a little contradictory, at least not very helpful.

C Lattice energy

The electrostatically defined lattice energy is a true dinosaur of chemical reasoning for the solid state, easily explained by one or two simple examples. For the case of rock salt, NaCl, the notion is that NaCl can be somewhat magically evaporated into isolated **ions** of Na^+ and Cl^-, as given in

$$NaCl_{(s)} \rightarrow Na^+_{(g)} + Cl^-_{(g)},$$

a rather bold assumption used to define an energetic reference point to which all other (often pretty advanced) calculations in the solid need to be compared. Nonetheless, one finds a rather weakly bonded NaCl **molecule** (not held together by ionic forces) in the gas phase with a Na–Cl distance of 2.36 A (Müller et al., 2021), so the reference point of ions being infinitely apart from each other is certainly a great idea (corresponding to $V \rightarrow \infty$ and $p \rightarrow 0$) but nonetheless physically questionable. It is of course possible to compare the different lattice energies of, say, NaCl and KCl, such that their **relative** sizes tell us something about their stabilities expressed through melting temperatures but the absolute size of the lattice energy lacks physical meaning, irrespective of how "accurately" it has been calculated. And still the electrostatic lattice energy (built upon Coulomb's law) will not go extinct, that is clear, but it will survive as a living fossil in the textbooks because it is so easy to understand. Fine accounts have been written (Greenwood, 1968).

The same lattice energy is even "more wrong" when used for a ternary phase such as $BaTiO_3$ because, once again, the notion exists that $BaTiO_3$ can be evaporated into **ions** of Ba^{2+}, Ti^{4+}, and O^{2-}. We have now reached the realm of pure science fiction, of course, simply because an isolated O^{2-} does **not** exist, it is thermodynamically unstable, hence the meaning of such lattice energy is somewhat ill-defined, to put it mildly; note that many anions only exist (in the solid) because they are stabilized by all the surrounding cations. Relative lattice-energy comparisons may be perfectly valid, of course. So, we may compare $BaTiO_3$ with $SrTiO_3$ in terms of lattice energy, that is possible.

And let us not forget that there is a difference between **lattice** energy and **Madelung** energy, sometimes (incorrectly so) neglected here and there. To calculate the electrostatically defined lattice energy, the entire electrostatic energetics is summed up for the crystal **including** the short-distance (Pauli) repulsion term between all the ions; because only cations and anions come close to each other, this only affects those shortest interionic contacts. If one **excludes** this repulsion term on purpose, the hence defined Madelung energy is therefore a little more negative, on the order of 10%, than the lattice energy (Dronskowski, 2005).

D Quantities and entities

Without any doubt, quantum theory and quantum (or wave) mechanics belong to the greatest achievements of mankind. These parts of the natural sciences are a true bliss not only because they let us understand the atomistic world and then calculate the behavior of atoms and molecules and even larger systems up to the macroscopic human scale with a fascinating accuracy but they have also allowed us to sharpen our understanding of the world **per se**. For example, the behavior of entangled photons is clearly counterintuitive but it manifests the quantum foundations of reality, as weird as they may look at first sight. We still have no idea how quantum computers will change quantum chemistry but these developments will be exciting to watch.

And yet, chemically thinking **only** in terms of the Schrödinger equation and its equation relatives also may have drawbacks. In a spiteful **cartoon** of a computational scientist (mostly a theoretical physicist, but he or she can also stem from chemistry), the guy is sitting in the smallest thinkable box, also called **reductionism** (Primas, 1981) and judges whatever scientific result on the basis of whether or not it refers to what can be experimentally observed, a quantum-mechanical **observable quantity**, so to speak. This train of thought is not necessarily false, at least as long as the terms are clearly defined, in particular, **observable quantity**, but it may lead to infertile consequences. Here is such an idea heard over and over again at coffee breaks: for any **observable quantity** or simply **observable**, there is a self-adjoint (or Hermitian) **operator** which, by applying it to the Schrödinger equation, yields that **observable**. For example, the Hamilton operator yields the energy which can be measured (say, a few eV, thereby **quantified**). If there is no Hermitian operator, there is no observable, and "it" (whatever "it" stands for) cannot be measured. So, things that cannot be observed or measured do not exist, at least they do not exist in the minds of those people, end of story.[7]

This supposedly exact truth of quantum theory looks like a gross oversimplification of our world to me. First, a moment reflection reveals that any observable is ra-

---

7 Even today, some people are not aware of the fact that a wave function can be properly reconstructed from experiment, but that is another topic (Schwarz, 2006).

tionally deduced from whatever observation **only with the help of theory**. Second, one needs to **separate** the microscopic object (say, a molecule) for making it theoretically tractable, another huge intervention that will generate problems later in the analysis. If there is a problem, who is to blame? But let us get more practical:

Consider the simple example of benzene, C_6H_6, the prototype of an aromatic molecule, a liquid phase under standard conditions, having a very characteristic (aromatic, hence its name) smell. And now we replace one of the H atoms by a carboxyl group and arrive at benzoic acid ($C_6H_5CO_2H$), which appears as a solid-state material. The reader can be assured that the Schrödinger equation of these two molecules will look exactly the same, and the theorists will have a hard time – unless he or she has learned chemical thinking – to **really** extract the chemical difference from the numerical result. Why is that? The definition of a **functional group** such as the carboxyl group is a chemical invention, and while it does **not** translate into quantum physics, it **defines** the chemical behavior of benzoic acid being acidic, just like other acids having such a functional group. Sure, we may rely on the Born–Oppenheimer approximation, fix the nuclei, and then do two independent calculations, but how do we **identify** the carboxyl group? The reader can also be assured that there is no Hermitian operator for such functional groups, so what should we do about the functional group computationally? It is not an observable; hence, it does not exist?

The functional group CO_2H exists, of course, and it not only exists in benzoic acid, it also shows up in countless other organic acids,[8] one way to chemically identify so-called **entities**, not observable quantities. The carboxyl group is a chemical **entity**, the carbonyl group is another **entity**, the imine group is yet another **entity**, and there are plenty of other chemical **entities** introduced long ago to shape chemistry as an "independent" science, to make sense out of chemistry, successfully so. Chemistry is like that, and its complexity requires such kind of "fuzzy" logic to advance. And because the entity is perfectly valid and can be manipulated in the hands of a skillful chemist, one may use a Friedel–Crafts acylation reaction to turn benzene into benzoic acid. Yes, one entity can be changed into another entity. That being said, an entity is an object that can be fuzzily defined in the language of the corresponding science (here: chemistry), and it moreover can be experimentally studied (or "prepared," as the physicists usually coin it) for proper investigation. You may have guessed it already: the chemical bond is an **entity**, not a quantity, but it can be quantified after the entity has been skillfully formulated. Even though there is no Hermitian operator for chemical bonds, we can study them carefully and yield useful results, but only if we have properly formulated what we mean by that, for example, a two-electron single bond in CH_4 and SF_6 but not necessarily in NaCl and CaO.

If the quantity-entity antagonism, as we just derived it from (molecular) chemistry, does not convince you, here is another example. An atom consists of the nucleus

---

8 If you scroll back, the molecule shown in Figure 3.4 is like that.

and the electrons moving around it but there is no Hermitian operator whatsoever which is able to yield the nucleus as an observable quantity.[9] In case of doubt, try to imagine the physical unit of the nucleus. Pounds per square inch? Miles per gallon? Fahrenheit? I am joking, of course. But the atomic nucleus is a well-defined **entity**, and given proper instrumentation, it can also be studied and manipulated, too. So, Hahn and Strassmann (1939) managed to split the atomic nucleus in 1938 and also proved that because they recognized entities. By doing so they chemically redefined nuclear science through breaking the physical laws of that time (and smashing the tiny box mentioned at the beginning), thereby opening up a new area of nuclear physics, too. Think about **entities**, they are useful constructs, in particular, for chemistry and consider studying chemical philosophy (Schummer, 1998).

E The special role of the second period (i.e., the elements Li–Ne)

From Appendix A, the reader may recall that elements such as C, N, and O play a special role in chemistry in that they strictly follow the octet rule which is not obeyed by their higher homologues, though. And these light atoms also readily form double or triple bonds, unlike their higher homologues. And the elements of the second period (or first **long** period, meaning the same) are truly small atoms, much smaller than the higher homologues, and they are also quite electronegative, more electronegative than the heavier ones. Why is that?

Before we solve this puzzle, let us remind ourselves that the aforementioned light elements more or less **define** molecular organic chemistry, so despite their small number (we are talking about a handful of elements, not more, certainly not the left-over 100 or so which make up the entire periodic table) they are chemically **important**, so the chemists have condensed their chemical and bonding behavior into various rules (the octet rule being the most important one). That being said, chemical-bonding theory in textbooks is usually introduced using such small molecules made up of those light atoms, for good or at least understandable reasons. When going to heavier molecules composed of higher homologues, however, all those molecules then look like weird exceptions, so there are more exceptions than anticipated, simply because the chemists have decided to designate the chemistry of the true exceptions (C, N, O, etc.) as **normal** and the chemistry of all the other elements of the periodic table (which are clearly in the majority) as **exceptional** (Kutzelnigg, 1984). Funny, isn't it? It is a matter of chemical psychology and the dominance of organic chemistry, in addition to the importance of the entire biological world, so to speak.

But why are those light elements from the first long period so different? While this is quite easy to explain, it still hasn't found enough coverage generally, so let's

---

9 I owe this nice example to my good old friend Andrei L. Tchougréeff.

cover it here, a bit simplified. When the first **short** period has been completed and the [He] configuration $= 1s^2$ is accomplished, the $2s$ orbital will get filled starting with the lithium atom, later followed by the three $2p$ orbitals, so boron is having an electron configuration of $2s^2 2p^1$, carbon having $2s^2 p^2$, nitrogen having $2s^2 2p^3$, oxygen having $2s^2 2p^4$, and fluorine having $2s^2 2p^5$. Now, one needs to know that all those valence electrons are shielded from their nuclei by the inner core electrons ($1s^2$ in this case) but to a different extent. Because of their different topologies, orbitals with the same angular-momentum quantum numbers will shield their cousins quite well from the bare nuclear potential since their **shape** is similar, so s orbitals nicely shield s orbitals, p orbitals shield p orbitals, d orbitals shield d orbitals, and f orbitals shield f orbitals – but an s orbital will *not* sufficiently shield a p orbital, and an f orbital will not shield a d orbital, for example (Burns, 1964).

And then we eventually realize that someone "forgot" to establish the $1p$ orbital, so to speak, it simply isn't there given the quantum-mechanical laws applied to an atom (i.e., a spherical potential). Hence, while the $2s$ electrons are shielded by the inner $1s^2$ shell, this effect isn't there for the $2p$ orbitals and their electrons which, as a consequence, must see a **higher** nuclear charge than for the $2s$ case.[10] Because of that, the $2p$ orbital shrinks (so the atom as a whole gets smaller), its atomic eigenvalues get lower in energy (so incoming electrons will gain more energy, will get more attracted, so the atom becomes more electronegative), and $2s$ and $2p$ orbitals occupy about the same spatial region (allowing for sp mixing or "hybridization" as the physicists call it, see Appendix H) which eventually makes multiple bonding possible, say, in O_2 and N_2. Likewise, nature basically treats s and p electrons on equal footing as they reside in the same spatial region, and a simple octet approaching the [Ne] configuration for **both** s and p **together** will do. And this is only found for the second (or first long) period, hence its exceptional chemical bonding.

Beware of the missing $1p$ orbital! It explains why the second period is different and why organic chemistry (and also ourselves being composed of such molecules) does exist, thank God (Harris & Jones, 1979). It is also relevant for materials science, of course, by making the dielectric BeO different from all the other alkaline-earth oxides (Reitz et al., 2023). A similar effect also exists for the first d period, the elements for which the $3d$ orbitals are being filled (see Section 4.5). Because there is no $2d$ orbital (an atomic "weaving mistake," once again), the $3d$ are not sufficiently shielded and get more contracted, so these atoms yield magnetic or colored compounds, for example. And because there is no $3f$ orbital, the $4f$ elements (the lanthanides) have very contracted $4f$ orbitals which also makes them splendid atoms for applications in magne-

---

10 Alternatively one might say that, while $2s$ must be orthogonal to the inner $1s$ (which is filled), $2p$ does not have to be orthogonal to $1p$ because there is no $1p$. Of course, $2p$ must be orthogonal to $1s$ but still it is more contracted compared to $3p$ which must be orthogonal to a filled inner $2p$.

tism and light emittance with very sharp wavelengths. The effect has been given a name, it is sometimes called "primogenic repulsion" (Kaupp, 2007).

F Oxidation states (or numbers), ionic charges, and formal charges

I am sorry; this is another section about nomenclature. Most chemists know (or should know) that all three aforementioned terms **mean entirely different things**, so linguistic confusion and misunderstanding is guaranteed if one talks to people from other fields, people who are smart but lack chemical training. In order to not let such misunderstanding happen, let me briefly summarize what is meant by those chemical expressions.

The **oxidation state** (or number) is the totally **fictitious charge** of an atom in a molecule or a molecular ion assuming that all polar bonds (according to the electronegativity differences) are split as if they were totally ionic, the "ionic limit," so to speak. That sounds a bit abstract, so let us have a few examples. In sodium hydride (NaH), Na is less electronegative than H, so we formulate it as Na^+H^- or Na(+I) and H(–I) either using superscript Arabic or regular Roman numbers. Both ions then adopt noble-gas configurations, [Ne] for Na and [He] for H, and H turns into an anion, hence the chemists call it **hydride**. In hydrogen chloride (HCl), however, the EN difference lets us write H^+Cl^- or H(+I) and Cl(–I), very easy; now chlorine is the anion, hence **chloride**. In a simple molecular oxide such as carbon monoxide (CO), we would write the oxidation states (or numbers) as $C^{2+}O^{2-}$ or C(+II) and O(–II). And we also do the same for solid-state calcium oxide (or lime) (CaO), so here the oxidation states are $Ca^{2+}O^{2-}$ or Ca(+II) and O(–II). If there is an O–O single bond such as in H–O–O–H (hydrogen peroxide), the oxidation states can only read $(H^+)_2(O^-)_2$ or H(+I) and O(–I). And for MnO_2 we have $Mn^{4+}(O^{2-})_2$ or Mn(+IV) and O(–II); in $KMnO_4$ the oxidation states read $K^+Mn^{7+}(O^{2-})_4$, and even in polar covalent IF_7 we would write them as $I^{7+}(F^-)_7$, irrespective of the covalent character; chemists just want to balance the valence electrons assuming the ionic limit. It is easy, isn't it?

Now just imagine that we would want to calculate the lattice (or Madelung) energy of crystalline CaO, which **ionic charges** would we use? By something that might be called crystal-chemical tradition, chemists would usually take the oxidation states as ionic charges, so +2 for Ca and –2 for O, a highly simplified notion but such high charges seem plausible by **also** assuming the ionic limit. People really do this when calculating the lattice (or Madelung) energy, at least it **has** been done in the past. And let us not forget that in the first half of the twentieth century, there was hardly any alternative to this oversimplified ionic notion, so let us please be tolerant in the twenty-first century. A moment reflection will also yield that this notion might be perfectly fine for a supposedly ionic phase such as CaO but for a more molecular compound, say, solid Mn_2O_7, the real charge on Mn is certainly **not** +7 but much smaller, and it will also be smaller than normal for O, not –2, the reason being that highly unstable and explosive Mn_2O_7 is a

molecular compound,[11] so the existence as a molecule suggest covalency, not ionicity. And still, taking +7 and −2 for Mn and O (sometimes called "picture-book charges," from the German *Bilderbuchladungen*) seem plausible **if and only if** one sticks to the ionic limit. It is an enormous oversimplification that should not be forgotten.

For a better calculation, more realistic charges would be based on a quantum-mechanical calculation, either a partitioning of the wave function (say, Mulliken or Löwdin charges) or of the density (Bader charges); the latter is particularly problematic if atoms overlap (which they do). In any case, all such partitionings are somewhat **arbitrary** because it is physically **impossible** to measure a charge of an atom inside a molecule (since it is no observable although the charge of a simple or complex ion as a whole can be measured, of course), and one is well advised to compare calculated charges with other physical properties such as to derive **indirect** conclusions. To do that, the charges should be **chemically meaningful**, simply said. An atom inside a molecule is an entity, not an observable quantity, as you may guess.

Formal charges are yet another term, and it has **nothing** to do with the former charges. To calculate formal charges, one assumes **homolytic** bond breaking without considering electronegativity differences, so we strive for the "covalent limit." A simple example will do; we once again take the CO molecule, which has a $C{\equiv}O$ triple bond in-between and one nonbonded electron pair on both C and O, so we could write $|C{\equiv}O|$ for the Lewis-type electronic configuration, with 6 electrons shared between C and O, and 2×2 lone-pair electrons, in total 10 electrons; while CO does not violate the octet rule but has an octet for both C and O, it is not too stable and loves to be oxidized to the thermodynamic sink carbon dioxide, CO_2.

Partitioning the 6 shared electrons between C and O homolytically, 3 go to C, and 3 go to O, so the atomic electron count then is $2 + 3 = 5$ for C and $2 + 3 = 5$ for O. But since the neutral C atom has 4 valence electrons, it is now **formally** charged −1 by assuming the covalent limit. And since neutral O has 6 valence electrons, it has become **formally** charged as +1 within CO; of course, this runs against electronegativity but this is how chemists do it. And to indicate that they are not entirely serious, the chemists draw little squiggles around the charges, as shown in the upper left part of Figure A2.

The wealth of chemical experience shows that, given that there are a couple of alternative valence-bond electronic configurations for the same molecule, those which mirror the electronegativities are the ones closest to reality (or, more theoretically expressed, dominant in the resonance mixtures). In CO, the Lewis sketch shown on the upper left side of Figure A2 corresponds to this configuration (because only this one allows for a 1.13 Å triple bond), so we are already set. Nonetheless, this config-

---

11 The molecule consists of two MnO_4 tetrahedra which share a common corner O atom. In the crystal structure, all Mn_2O_7 molecules are packed in such a way as to allow for an approximate *fcc* packing of the O atoms, and the tetrahedral voids are filled in an ordered way by Mn. This molecule once was important to me.

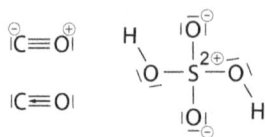

Figure A2: Formal charges in the Lewis formula of molecular carbon monoxide (top left) and an alternative Lewis formula involving a so-called "dative" bond (bottom left). The improved Lewis formula of sulfuric acid involving only single bonds but with formal charges is shown on the right.

uration with C being negative and O being positive (**opposed** to the EN difference) lets us expect a molecular dipole moment with its arrow pointing from O to C, from positive to the negative, and this is – surprisingly – also found in experiment: the dipole moment is nearly zero, **suggests** C to be charged **slightly** negative, the O atom **slightly** positive, so the formal charges of CO do make sense chemically, as weird as that may seem, and the small dipole moment of about 0.11 Debye may be easily understood from the quantum-chemical analysis (Frenking et al., 2007).[12] For completeness we also draw another Lewis formula on the lower left, this time involving a "dative" bond (Nandi & Kozuch, 2020) which already indicates – without formal charges – that electron density is donated from O to C, in harmony with the experimental dipole moment. For whatever reason, this kind of Lewis formulation has not found widespread use.

The recipe of drawing formal charges may also serve us well to come up with an improved Lewis formula for sulfuric acid, depicted on the right of Figure A2, to be compared with Figure A1. In the new representation, there are only single bonds between sulfur and oxygen; sulfur now has an electron octet, but two O atoms are formally charged –1 whereas the sulfur atom is charged +2. This more ionic picture indicates that all the S–O bonds are more similar to each other (which is true) but it overstresses the difference in charge transfer of the S–O bonds (which is not true). It **is** difficult to draw a realistic picture of quantum chemistry using classical means.

G Chemistry of special relativity

One would not necessarily expect an impact of Einstein's theory of special relativity (Einstein, 1905) on chemistry but for quite a few (important) elements the oversimplification of using a nonrelativistic wave equation such as the one by Schrödinger (Schrödinger, 1926a; Schrödinger, 1926b) for all elements of the periodic table simply yields an incorrect description of reality here and there. In fact, only two years after Schrödinger's enormous achievement, Dirac came up with a relativistic version of the

---

12 The situation is more complex: C is slightly **positively** charged but one should not naively use such charges to calculate dipole moments because the **topography** of the charge distribution is important.

Schrödinger equation, the so-called Dirac equation (Dirac, 1928), which not only forms the basis of relativistic quantum chemistry but also predicts the existence of antiparticles, that is, the same elementary particles but with different plus–minus charges. Dirac, by the way, grossly underestimated the consequences of his equation by later falsely writing that "relativistic effects are of no importance in the consideration of the atomic or molecular structure and ordinary chemical reactions." Here, ingenious Dirac was pretty wrong,[13] probably because he failed to discuss his ideas with an experienced inorganic chemist.

We do not have enough time to really delve into relativistic quantum chemistry (Dyall & Fægri, 2007; Schwarz, 2010) which comes in different mathematical representations, and they **are** difficult to comprehend. Alternatively, we may look at relativity from the perspective of the Schrödinger equation, and then the following **relativistic effects** (Pyykkö, 1988) show up as perturbations, to be added to the Schrödinger picture.

Because there is a non-zero probability for an s electron to go through the atomic nucleus, the $1s$ electron of a heavy atom (say, Hg with $Z = 80$) approaches relativistic speed, and hence, it will undergo a relativistic mass increase such that the orbital it belongs to **shrinks** in diameter (or contracts, another way of saying it). This effect is present for all s electrons (and the similar $p_{1/2}$ electrons, see below), so it also applies to some of the **valence** electrons but Dirac forgot to consider that. Second, as a consequence of Heisenberg's uncertainty principle, the mass and the charge of an s electron close to the nucleus do not fully coincide in space, so this is dubbed *Zitterbewegung* (German for "trembling motion") as if the electron's mass and charge were smeared out a little; the effect is small and slightly counteracts the former orbital contraction. Third, the phenomenon widely known as spin-orbit coupling going quadratic with the nuclear charge (hence, $\sim Z^2$) is entirely relativistic, leading to new orbitals (so-called spinors) with different quantum number $\vec{j} = \vec{l} + \vec{s}$ and $j = l \pm 1/2$ for p, d, and f orbitals. For example, the formerly degenerate set of three p orbitals change into one spherically shaped $p_{1/2}$ and two $p_{3/2}$, and for the five d orbitals they are two $d_{3/2}$ and three $d_{5/2}$. These orbitals are also nodeless, not like the "normal" ones. Fourth, the three effects superimpose, in particular as regards the contracted $s_{1/2}$ and $p_{1/2}$ spherical spinors shielding the nuclear Coulomb attraction, and this leads to **less** contracted and energetically higher lying d and f levels, the most important consequence.

Compared to the nonexisting nonrelativistic world (let us not forget that our world **is** relativistic, the reference world does **not** exist but is a true *Gedankenexperiment*), one often finds bond shortening when heavy atoms are involved (Pyykkö, 1988). For example, a molecule such as TlH is slightly smaller than in a nonrelativistic world, similar to the effect found in molecular PbO. Even the lanthanide contraction carries a little (10%) relativistic character. In contrast, one finds an enlargement for

---

13 To rephrase Lichtenberg: If you don't understand anything but physics, you do not really understand it either.

the 5*f* elements due to an indirect effect. The higher homologue of Cu, Ag, and Au dubbed roentgenium (Rg), synthesized in 1994, was theoretically calculated as being **smaller** than all of its lighter cousins (Fricke, 1975). For all the heavy main-group elements, the 6*s* orbitals are strongly contracted, corresponding to an extra stabilization of the $6s^2$ configuration, so these two electrons drop in energy and are less available for chemical bonding. Hence, the preferred oxidation number of thallium is +I but not +III, opposed to indium. Also, Pb^{2+} but not Pb^{4+} is a good oxidation state (unlike in tin), and the lead battery "lives" from that, without relativity its voltage would be much smaller. And the story continues for main-group V with Bi^{3+} (preferred) instead of Bi^{5+} (not preferred). This is where the "inert-pair" or "lone-pair" effect for the heaviest main-group elements **really** originates from.

All the direct and indirect effects somehow culminate for the gold and mercury atoms, making them truly special, Au in particular. Because of relativity, Au looks golden but not silvery (the optical gap shrinks), and it is also a pseudohalogen atom, similar to the heavier halogens in terms of the electron affinity and electronegativity (close to iodine) and also the bond energy of Au_2 (close to Br_2). Hence, the existence of a **semiconducting** but not metallic intermetallic compound dubbed $CsAu = Cs^+Au^-$ (see Section 4.5) should not be too surprising, Einstein must be blamed (I am smiling).

H Hybridization and orbital mixing

I find it quite funny, in a sense, that entirely different nomenclatures have been developed in quantum-physical and quantum-chemical theory, in particular, as regards the use of the word **hybridization** which carries dissimilar meanings for physicists and chemists. Imagine you look at the DOS of silicon dioxide in a paper that came out in a well-respected physics journal, and then it is highly likely that the authors (mostly from physics) realize that both Si and O contribute to the occupied levels below the Fermi level, easily seen from a local DOS. And while the DOS does not carry any bonding information (remember that the orbital phases are missing), the energetic proximity of the Si 3*p* levels and the O 2*p* levels and their similar dispersion will let the authors conclude "that silicon and oxygen states **hybridize** in the valence region" or something similar to that.[14] With the word **hybridize** the authors simply want to indicate that the atomic orbitals of Si and O do **interact** or **mix**, as the chemists would call it. And it is true, irrespective of whether one calls it hybridize or interact or mix. So, the physicist's hybridization is the chemist's **orbital mixing between atoms**, nothing to worry about.

---

14 Please note that they often talk about **states** (which imply many-body theory) although one-electron **levels** are meant, but this sloppiness also translates into the naming of the DOS which should better be called DOL (for density of levels). Nobody cares and nobody will ever change that.

When chemists are talking about hybridization, however, they allude to Pauling's ingenious idea to simplify the quantum-chemical treatment of small molecules such as methane, CH_4 (covered in Appendix A). In the early days of quantum chemistry when everything was calculated by hand, people had to be smart to speed up things. So, instead of calculating the overlap of the one $2s$ and the three $2p$ orbitals of C with the four $1s$ orbitals of the four H, one may "hybridize" (= unitarily transform) the $2s$ and $2p$ orbitals of C into a fully equivalent set of four sp^3 hybrid orbitals which have the magnificent property of pointing into the four corners of a tetrahedron, the shape of the methane molecule, and then one only needs to calculate the overlap of a carbon sp^3 orbital with a hydrogen $1s$ orbital, end of story. Pauling was brilliant indeed (Pauling, 1960).[15] No chemistry or physics whatsoever is involved in this hybridization, a simple change in representation or orbital topology.

Admittedly, some (organic) textbooks then arrive at the terribly wrong conclusion that CH_4 is tetrahedrally shaped "because the C atom is sp^3-hybridized." And some physics texts also say that the C atom in diamond is sp^3-hybridized (which is at least misleading) and that even the Si atom in crystalline silicon is sp^3-hybridized (which is clearly wrong). For example, the photoelectron spectrum of methane, CH_4, shows an a_1 ($2s$-like) and a broader t_2 ($2p$-like) level, not something resembling sp^3 – remember that hybridizing $2s$ and $2p$ was only done for reasons of computational efficiency in valence-bond theory. In molecular-orbital theory, a regular, nonhybridized basis will do fine.

So why is methane tetrahedral? There will be four strong single bonds between C and H making the C atom approach an electron octet, so these C–H = 1.09 Å two-center two-electron bonds lower the energy to begin with. Likewise, the H atoms have the [He] configuration, so there is no interaction between the H unless they come too close to each other, and then they will feel some Pauli repulsion between the filled $1s^2$ atomic orbitals. For attractive C–H and nonattractive, possibly repulsive H–H interactions, the tetrahedron is nature's best choice by simple steric reasons.[16] Alternative square-planar structures, for example, require special electron fillings, namely the d^8 configuration seen in divalent platinum.

Likewise, tetrahedral C in the diamond structure is an excellent choice for four two-center two-electron bonds, giving a total energy that is extremely close to the true ground state, graphite, even analytically (Popov et al., 2019). And in crystalline silicon, the diamond type also allows for an optimum Si–Si bond. As alluded to before, the sp^3 hybridization of Si is clearly wrong due to the bad overlap of $3s$ and $3p$ levels, the Si atom not belonging to the first long period (see Appendix E), so it took the chemists

---

15 Pauling knew that in valence-bond theory only one-center and two-center overlap interactions are allowed; hence, one must choose a basis that guarantees the correctly assumed result.

16 It has been shown that CH_4 stays tetrahedral even when mixing between $2s$ and $2p$ is artificially suppressed on purpose, by clever theory (Ahlrichs, 1980), so nature does not need sp^3 for tetrahedral geometry.

quite a while until a Si=Si double bond was eventually made. Coming back to structural questions of elemental silicon, the highly simplifying notion that tetrahedral Si is sp^3-hybridized and trigonal-prismatic one is sp^2 is incredibly widespread (within physics in particular), so widespread actually that people take it for granted. It remains incorrect, though, because in crystalline silicon, the Si–Si bond is dominated by the $3p$ orbital.

By the way, to allow for the proper treatment of octahedral molecular shapes, Pauling then went on and invented the d^2sp^3 hybrid orbital. This species only has relevance in outdated textbooks; it went extinct in today's chemistry because it simply does not work.

I Superconductivity and chemistry

Solid-state chemists usually feel a bit uneasy about superconductivity ever since its discovery in 1911 because this very macroscopic quantum effect does not relate very well to chemical thinking, with serious consequences for research until today. At a given (very) low critical temperature T_c, some (possibly all, nobody knows) materials lose their electrical DC resistance and arrive at a **zero** value such that an "eternal" electric current may remain, and there is an onset of ideal diamagnetism, the so-called Meißner–Ochsenfeld effect (Ashcroft & Mermin, 1976). It took the physics community quite a while to understand that an entirely different kind of matter is involved, and the carriers of superconductivity follow Bose–Einstein, not Fermi–Dirac statistics like in regular matter. Even though the **bosons** (**paired** electrons, so $S = 0$ for the total spin quantum number) making up the superconducting current are in the clear minority, they short-circuit the normal (ohmic) current, hence the material becomes superconducting as a whole.

Coupling fermionic electrons to pairs of electrons (then becoming bosons typically called **Cooper pairs** in this context) needs a special mechanism, for example, one operating through collective vibrations (phonons), and this so-called BCS theory turned out to be exceedingly fruitful, was worth a Nobel prize (Bardeen et al., 1957), and it is the **de facto** standard of superconducting reasoning, confirmed over and over again by the isotope effect.[17] Other, more local, more **electronic** mechanisms are thinkable, too. Despite its success, the BCS idea operates in reciprocal space, the electron pairs are intellectually inapprehensible (what is a boson, really?), and chemical thinking ultimately connected with fermionic reasoning (orbitals, levels, bands, etc.) collapses for superconductivity. And **this** is the reason why nobody (chemists, physicists, material scientists, whoever) is able to write down the chemical formula of a fine new superconductor

---

17 Due to the phononic coupling, the coherence length (= coupling distance of the paired electrons) may be **thousands** of Ångstroms since the time scales of phonons (slow) and electrons (fast) differ a lot.

with a transition temperature of 305 K at standard pressure, nobody has a clue, in sharp contrast to predicting yet another semiconductor with a bandgap smaller than the one of, say, GaN (GaP, of course, one does not need DFT for that).

Likewise, existing superconductors are chemically so **diverse** that it is impossible to see the forest for the trees. The first elemental one was Hg, another one is Pb, Nb_3Sn is a common intermetallic superconductor, the high-temperature superconductor revolution started with the $La_{1.85}Ba_{0.15}CuO_4$ ceramic and $T_c = 35$ K (Bednorz & Müller, 1986), followed by $YBa_2Cu_3O_7$ ($T_c = 93$ K, suitable for liquid nitrogen), but fulleride salts such as Rb_3C_{60} ($T_c = 29$ K) and organic charge-transfer complexes (Saito & Yoshida, 2011) are superconductors as well, even iron-containing arsenides such as $Ba_{1-x}K_xFe_2As_2$ at 38 K (Rotter et al., 2008). What do they have in common, chemically? Our astonishing **inability** to recognize the low-temperature property from structure and composition is perfectly demonstrated from the MgB_2 story: although known since 1912, this Zintl phase – hence, $Mg^{2+}(B^-)_2$ with Mg^{2+} ions between graphene-like B sheets – superconducts at 39 K, but nobody had bothered making the measurement until 2001, presumably a random discovery (Nagamatsu et al., 2001).

And yet, there **are** fingerprints for superconductivity, the Bose–Einstein world does "mirror" into our Fermi–Dirac world. For example, a good metallic conductor (Cu, Ag, Au) is not a superconductor; one needs a proper valence-electron concentration (Matthias, 1955); a high $DOS(\varepsilon_F)$ will induce a high transition temperature; ferromagnetism (hence, Fe) competes with superconductivity (iron arsenides being exceptions); extreme pressures (in hydrides, hundreds of GPa) induces superconductivity; and so forth, all findings not very chemical in style.

A **chemical** picture was first developed by Ogg who identified electron pairs with opposing spins inducing diamagnetism and superconductivity (Ogg, 1946).[18] Paired spins in atomic or molecular orbitals resemble chemical thinking, of course, and we have witnessed complex anions with weird electron fillings already. It turns out that there is an entire **class** of two-dimensional metals such as $Ln_2X_2C_2$ (with $Ln = 4f$ metal and $X = $ halogen) which are superconductors, and their behavior may be chemically tuned, for example, by substitution, by intercalation, and by isotope replacement. And since their ionic electron count is $(Ln^{3+})_2(X^-)_2C_2^{4-}$, the structurally sandwiched carbon dimer corresponds to N_2^{2-} or O_2, so the relevant electrons are likely to be paired in the $1\pi_g*$ molecular orbital, see Figure 4.54, overlapping with empty $5d$ levels. It can be shown, both experimentally and theoretically, that the source of superconductivity in these compounds must be pretty **local**, and the corresponding band structure includes both regions of **flat** bands (hence, localized, C_2-related) at some regions in reciprocal space but **steep** bands at others (delocalized, related to Ln–C covalency). And

---

18 Ogg's fascinating findings from quickly quenched alkali metal solutions in liquid ammonia could not be reproduced, however. He later became depressive and committed suicide.

this mysterious **chemical** fingerprint also shows up for superconducting MgB_2 and superconducting high-pressure Te (Simon, 2015).

Whether or not such fingerprints will eventually be explained in full by a chemical theory of superconductivity is unclear at the moment. In terms of theoretical support, however, one also needs proper computational methods, in particular, approaches being able to correctly calculate the BCS ground state. Fortunately enough, some progress has been achieved, and one may start with the effective one-particle (DFT) picture from plane waves, project that information to atomic orbitals, and then derive an electronic structure from that which allows one to theoretically model the superconducting (BCS) and also spin-liquid (resonating valence bond; RVB) phases (Plekhanov et al., 2020; Tchougréeff et al., 2021).

References

Ahlrichs, R. (1980). Gillespie- und Pauling-Modell – ein Vergleich, *Chem. unserer Zeit*. **14**, 18–24.

Albright, T. A., Burdett, J. K., & Whangbo, M.-H. (2013). *Orbital Interactions in Chemistry*, 2nd ed., John Wiley & Sons, Hoboken, New Jersey.

Andersen, O. K. (1975). Linear Methods in Band Theory, *Phys. Rev. B*. **12**, 3060–3083.

Andersen, O. K. & Jepsen, O. (1984). Explicit, First-Principles Tight-Binding Theory, *Phys. Rev. Lett.* **53**, 2571–2574.

von Appen, J., Lumey, M.-W., & Dronskowski, R. (2006). Mysterious Platinum Nitride, *Angew. Chem. Int. Ed.* **45**, 4365–4368.

von Appen, J., Eck, B., & Dronskowski, R. (2010). A Density-Functional Study of the Phase Diagram of Cementite-Type $(Fe,Mn)_3C$ at Absolute Zero Temperature, *J. Comput. Chem.* **31**, 2620–2627.

von Appen, J. & Dronskowski, R. (2011). Carbon-Induced Ordering in Manganese-Rich Austenite – A Density-Functional Total-Energy and Chemical-Bonding Study, *Steel Res. Int.* **82**, 101–107.

von Appen, J., Dronskowski, R., Chakrabarty, A., Hickel, T., Spatschek, R., & Neugebauer, J. (2014). Impact of Mn on the Solution Enthalpy of Hydrogen in Austenitic Fe-Mn Alloys: A First-Principles Study, *J. Comput. Chem.* **35**, 2239–2244.

van Arkel, A. E. (1956). *Molecules and Crystals in Inorganic Chemistry*, 2nd ed., Butterworth, London.

Ashcroft, N. W. & Mermin, N. D. (1976). *Solid State Physics*, Holt, Rinehart & Winston, New York.

Bachhuber, F., von Appen, J., Dronskowski, R., Schmidt, P., Nilges, T., Pfitzner, A., & Weihrich, R. (2014). The Extended Stability Range of Phosphorus Allotropes, *Angew. Chem. Int. Ed.* **53**, 11629–11633.

Bader, R. F. W. (1990). *Atoms in Molecules – A Quantum Theory*, Clarendon Press, Oxford.

Bardeen, J., Cooper, L. N., & Schrieffer, J. R. (1957). Theory of Superconductivity, *Phys. Rev.* **108**, 1175–1204.

Bartlett, N. (1962). Xenon Hexafluoroplatinate(V) $Xe^+[PtF_6]^-$, *Proc. Chem. Soc.* **6**, 218.

Batlogg, B., Kaldis, E., Schlegel, A., & Wachter, P. (1976). Electronic Structure of Sm Monochalcogenides, *Phys. Rev. B*. **14**, 5503–5514.

Becke, A. D. & Edgecombe, K. E. (1990). A Simple Measure of Electron Localization in Atomic and Molecular Systems, *J. Chem. Phys.* **92**, 5397–5403.

Bednorz, J. G. & Müller, K. A. (1986). Possible High T_c Superconductivity in the Ba–La–Cu–O System, *Z. Phys. B*. **64**, 189–193.

Benz, S., Missong, R., Ogutu, G., Stoffel, R. P., Englert, U., Torii, S., Miao, P., Kamiyama, T., & Dronskowski, R. (2019). Ammonothermal Synthesis, X-ray and Time-of-Flight Neutron Crystal-Structure Determination, and Vibrational Properties of Barium Guanidinate, $Ba(CN_3H_4)_2$, *Chem. Open*. **8**, 327–332.

Benz, S., Chen, D., Möller, A., Hofmann, M., Schnieders, D., & Dronskowski, R. (2022). The Crystal Structure of Carbonic Acid, *Inorganics*. **10**, 132.

Bernal, J. D. & Fowler, R. H. (1933). A Theory of Water and Ionic Solution, with Particular Reference to Hydrogen and Hydroxyl Ions, *J. Chem. Phys.* **1**, 515–548.

Bickelhaupt, F. M., Radius, U., Ehlers, A. W., Hoffmann, R., & Baerends, E. J. (1998). Might BF and BNR_2 Be Alternatives to CO? A Theoretical Quest for New Ligands in Organometallic Chemistry, *New J. Chem.* **22**, 1–3.

Bielec, P., Nelson, R., Stoffel, R. P., Eisenburger, L., Günther, D., Henß, A.-K., Wright, J. P., Oeckler, O., Dronskowski, R., & Schnick, W. (2019). Cationic Pb_2 Dumbbells Stabilized in the Highly Covalent Lead Nitridosilicate $Pb_2Si_5N_8$, *Angew. Chem. Int. Ed.* **58**, 1432–1436.

Biermann, S. & Lichtenstein, A. (2017). Many Body Perturbation Theory, Dynamical Mean Field Theory and All That, *Handbook of Solid State Chemistry* (edited by Dronskowski, R., Kikkawa, S., & Stein, A.), volume 5, p. 119–157, Wiley-VCH, Weinheim, New York.

Binnewies, M., Glaum, R., Schmidt, M., & Schmidt, P. (2012). *Chemical Vapor Transport Reactions*, De Gruyter, Berlin, Boston.

https://doi.org/10.1515/9783111167213-007

Bloch, F. (1928). Über die Quantenmechanik der Elektronen in Kristallgittern, *Z. Phys.* **52**, 555–600.

Blöchl, P. E. (1994). Projector Augmented-wave Method, *Phys. Rev. B.* **50**, 17953–17979.

Blöchl, P. E., Jepsen, O., & Andersen, O. K. (1994). Improved Tetrahedron Method for Brillouin-Zone Integrations, *Phys. Rev. B.* **49**, 16223–16233.

Borden, W. T., Hoffmann, R., Stuyver, T., & Chen, B. (2017). Dioxygen: What Makes This Triplet Diradical Kinetically Persistent?, *J. Am. Chem. Soc.* **139**, 9010–9018.

Boyko, T. D., Green, R. J., Dronskowski, R., & Moewes, A. (2013). Electronic Band Gap Reduction in Manganese Carbodiimide: $MnNCN$, *J. Phys. Chem. C.* **117**, 12754–12761.

Bredow, T., Lumey, M.-W., Dronskowski, R., Schilling, H., Pickardt, J., & Lerch, M. (2006). Structure and Stability of TaON Polymorphs, *Z. Anorg. Allg. Chem.* **632**, 1157–1162.

Bredow, T., Dronskowski, R., Ebert, H., & Jug, K. (2009). Theory and Computer Simulation of Perfect and Defective Solids, *Progr. Solid State Chem.* **37**, 70–80.

Bredow, T. & Jug, K. (2017). Semiempirical Molecular Orbital Methods, *Handbook of Solid State Chemistry* (edited by Dronskowski, R., Kikkawa, S., & Stein, A.), volume 5, p. 159–202, Wiley-VCH, Weinheim, New York.

Burdett, J. K. (1995). *Chemical Bonding in Solids*, Oxford University Press.

Burdett, J. K. (1997). *Chemical Bonds – A Dialog*, Wiley, Chichester.

Burdett, J. K. & McCormick, T. A. (1998). Electron Localization in Molecules and Solids: The Meaning of ELF, *J. Phys. Chem. A.* **102**, 6366–6372.

Burns, G. (1964). Atomic Shielding Parameters, *J. Chem. Phys.* **41**, 1521–1522.

Catlow, C. R. A., Buckeridge, J., Farrow, M. R., Logsdail, A. J., & Sokol, A. A. (2017). Quantum Mechanical/ Molecular Mechanical (QM/MM) Approaches, *Handbook of Solid State Chemistry* (edited by Dronskowski, R., Kikkawa, S., & Stein, A.), volume 5, p. 647–680, Wiley-VCH, Weinheim, New York.

Chadi, D. J. & Cohen, M. L. (1973). Special Points in the Brillouin Zone, *Phys. Rev. B.* **8**, 5747–5753.

Chadi, D. J. (1977). Localized-orbital Description of Wave Functions and Energy Bands in Semiconductors, *Phys. Rev. B.* **16**, 3572–3578.

Chen, Z., Löber, M., Rokicińska, A., Ma, Z., Chen, J., Kuśtrowski, P., Meyer, H.-J., Dronskowski, R., & Slabon, A. (2020). Increased Photocurrent of $CuWO_4$ Photoanodes by Modification with the Oxide Carbodiimide $Sn_2O(NCN)$, *Dalton Trans.* **49**, 3450–3456.

Chen, D., Wang, Y., & Dronskowski, R. (2023). Computational Design and Theoretical Properties of WC_3N_6, an H-Free Melaminate and Potential Multifunctional Material, *J. Am. Chem. Soc.* **145**, 6986–6993.

Chivers, T. & Laitinen, R. S. (2015). Tellurium: A Maverick Among the Chalcogens, *Chem. Soc. Rev.* **44**, 1725–1739.

Claassen, H. H., Selig, H., & Malm, J. G. (1962). Xenon Tetrafluoride, *J. Am. Chem. Soc.* **84**, 3593.

Clark, W. P., Steinberg, S., Dronskowski, R., McCammon, C., Kupenko, I., Bykov, M., Dubrovinsky, L., Akselrud, L. G., Schwarz, U., & Niewa, R. (2017). High-Pressure NiAs-Type Modification of FeN, *Angew. Chem. Int. Ed.* **56**, 7302–7306.

Coffey, P. & Jug, K. (1974). A Pedagogic Approach to Configuration Interaction, *J. Chem. Educ.* **51**, 252–253.

Condon, E. U. (1927). Wave Mechanics and the Normal State of the Hydrogen Molecule, *Proc. Nat. Acad. Sci. USA.* **13**, 466–470.

Corkett, A. J. & Dronskowski, R. (2019). A New Tilt and an Old Twist on the Nickel Arsenide Structure-Type: Synthesis and Characterisation of the Quaternary Transition-metal Cyanamides $A_2MnSn_2(NCN)_6$ (A = Li and Na), *Dalton Trans.* **48**, 15029–15035.

Corkett, A. J., Chen, Z., Bogdanovski, D., Slabon, A., & Dronskowski, R. (2019). Band Gap Tuning in Bismuth Oxide Carbodiimide Bi_2O_2NCN, *Inorg. Chem.* **58**, 6467–6473.

Coulson, C. A. (1964). The Nature of the Bonding in Xenon Fluorides and Related Molecules, *J. Chem. Soc.* 1442–1454.

Crowhurst, J. C., Goncharov, A. F., Sadigh, B., Evans, C. L., Morrall, P. G., Ferreira, J. L., & Nelson, A. J. (2006). Synthesis and Characterization of the Nitrides of Platinum and Iridium, *Science*. **311**, 1275–1278.

Decker, A., Landrum, G. A., & Dronskowski, R. (2002a). Structural and Electronic Peierls Distortions in the Elements (A): The Crystal Structure of Tellurium, *Z. Anorg. Allg. Chem.* **628**, 295–302.

Decker, A., Landrum, G. A., & Dronskowski, R. (2002b). Structural and Electronic Peierls Distortions in the Elements (B): The Antiferromagnetism of Chromium, *Z. Anorg. Allg. Chem.* **628**, 303–309.

Deringer, V. L., Tchougréeff, A. L., & Dronskowski, R. (2011). Crystal Orbital Hamilton Population (COHP) Analysis as Projected from Plane-Wave Basis Sets, *J. Phys. Chem. A.* **115**, 5461–5466.

Deringer, V. L., Hoepfner, V., & Dronskowski, R. (2012a). Accurate Hydrogen Positions in Organic Crystals: Assessing a Quantum-Chemical Aide, *Cryst. Growth Design.* **12**, 1014–1021.

Deringer, V. L., Lumeij, M., & Dronskowski, R. (2012b). Ab Initio Modeling of α-GeTe(111) Surfaces, *J. Phys. Chem. C.* **116**, 15801–15811.

Deringer, V. L., Lumeij, M., Stoffel, R. P., & Dronskowski, R. (2013). Mechanisms of Atomic Motion Through Crystalline GeTe, *Chem. Mater.* **25**, 2220–2226.

Deringer, V. L. & Dronskowski, R. (2013). Computational Methods for Solids, in *Comprehensive Inorganic Chemistry II* Vol. 9 (edited by Reedijk, J. & Poeppelmeier, K.), p. 59–87, Elsevier.

Deringer, V. L. & Dronskowski, R. (2014). Pauling's Third Rule Beyond the Bulk: Chemical Bonding at Quartz-Type GeO_2 Surfaces, *Chem. Sci.* **5**, 894–903.

Deringer, V. L. (2014). *Quantenchemische Modellierung und Bindungsanalyse komplexer Feststoffe*, Dissertation, RWTH Aachen University.

Deringer, V. L., Englert, U., & Dronskowski, R. (2014a). Covalency of Hydrogen Bonds in Solids Revisited, *Chem. Commun.* **50**, 11547–11549.

Deringer, V. L., Zhang, W., Lumeij, M., Maintz, S., Wuttig, M., Mazzarello, R., & Dronskowski, R. (2014b). Bonding Nature of Local Structural Motifs in Amorphous GeTe, *Angew. Chem. Int. Ed.* **53**, 10817–10820.

Deringer, V. L., Dronskowski, R., & Wuttig, M. (2015a). Microscopic Complexity in Phase-Change Materials and its Role for Applications, *Adv. Funct. Mater.* **25**, 6343–6359.

Deringer, V. L., Zhang, W., Rausch, P., Mazzarello, R., Dronskowski, R., & Wuttig, M. (2015b). A Chemical Link Between Ge–Sb–Te and In–Sb–Te Phase-Change Materials, *J. Mater. Chem. C.* **3**, 9519–9523.

Deringer, V. L., Stoffel, R. P., Wuttig, M., & Dronskowski, R. (2015c). Vibrational Properties and Bonding Nature of Sb_2Se_3 and Their Implications for Chalcogenide Materials, *Chem. Sci.* **6**, 5255–5262.

Deringer, V. L., Englert, U., & Dronskowski, R. (2016). Nature, Strength, and Cooperativity of the Hydrogen-Bonding Network in α-Chitin, *Biomacromol.* **17**, 996–1003.

Dierkes, H., Möller, A., & Dronskowski, R. (2013). Tribosynthesis of $Fe_{3-x}Mn_xC$ Phases *via* Mechanical Alloying and Annealing, *Z. Naturforsch.* **68b**, 1180–1184.

Dierkes, H. & Dronskowski, R. (2014). High-Resolution Powder Neutron Diffraction on Mn_3C, *Z. Anorg. Allg. Chem.* **640**, 3148–3152.

Dierkes, H., van Leusen, J., Bogdanovski, D., & Dronskowski, R. (2017). Synthesis, Crystal Structure, Magnetic Properties and Stability of the Manganese-Rich "Mn_3AlC" Kappa Phase, *Inorg. Chem.* **56**, 1045–1048.

Dinnebier, R. E., Etter, M., & Runcevski, T. (2017). Laboratory and Synchrotron Powder Diffraction, *Handbook of Solid State Chemistry* (edited by Dronskowski, R., Kikkawa, S., & Stein, A.), volume 3, p. 29–75, Wiley-VCH, Weinheim, New York.

Dirac, P. A. M. (1928). The Quantum Theory of the Electron, *Proc. Roy. Soc. Lond. A.* **117**, 610–624.

Djurovic, D., Hallstedt, B., von Appen, J., & Dronskowski, R. (2011). Thermodynamic Assessment of the Fe–Mn–C system, *CALPHAD.* **35**, 479–491.

Dolabdjian, K., Görne, A. L., Dronskowski, R., & Meyer, H.-J. (2018). Tin(II) Oxide Carbodiimide and Its Relationship to SnO, *Dalton Trans.* **47**, 13378–13383.

Dong, X., Oganov, A. R., Goncharov, A. F., Stavrou, E., Lobanov, S., Saleh, G., Qian, G.-R., Zhu, Q., Gatti, C., Deringer, V. L., Dronskowski, R., Zhou, X.-F., Prakapenka, V. B., Konôpková, Z., Popov, I. A., Boldyrev, A. I., & Wang, H.-T. (2017). A Stable Compound of Helium and Sodium at High Pressure, *Nat. Chem.* **9**, 440–445.

Dronskowski, R. & Blöchl, P. E. (1993). Crystal Orbital Hamilton Populations (COHP). Energy-Resolved Visualization of Chemical Bonding in Solids Based on Density-Functional Calculations, *J. Phys. Chem.* **97**, 8617–8624.

Dronskowski, R. (1995). Synthesis, Structure, and Decay of In_4Br_7, *Angew. Chem. Int. Ed. Engl.* **34**, 1126–1128.

Dronskowski, R., Korczak, K., Lueken, H., & Jung, W. (2002). Chemically Tuning Between Ferromagnetism and Antiferromagnetism by Combining Theory and Synthesis in Iron/Manganese Rhodium Borides, *Angew. Chem. Int. Ed.* **41**, 2528–2532.

Dronskowski, R. (2005). *Computational Chemistry of Solid State Materials*, Wiley-VCH, Weinheim, New York.

Dronskowski, R., Kikkawa, S., & Stein, A. (Eds.) (2017). *Handbook of Solid State Chemistry*, Volumes 1–6, Wiley-VCH, Weinheim, New York.

Dunitz, J. D. & Gavezzotti, A. (2012). Supramolecular Synthons: Validation and Ranking of Intermolecular Interaction Energies, *Cryst. Growth Des.* **12**, 5873–5877.

Dyall, K. G. & Fægri, K., Jr. (2007). *Introduction to Relativistic Quantum Chemistry*, Oxford University Press, Oxford.

Edmiston, C. & Ruedenberg, K. (1963). Localized Atomic and Molecular Orbitals, *Rev. Mod. Phys.* **35**, 457–464.

Eguía-Barrio, A., Castillo-Martínez, E., Liu, X., Dronskowski, R., Armand, M., & Rojo, T. (2016). Carbodiimides: New Materials Applied as Anode Electrodes for Sodium and Lithium Ion Batteries, *J. Mater. Chem. A* **4**, 1608–1611.

Eickmeier, K., Fries, K. S., Gladisch, F. C., Dronskowski, R., & Steinberg, S. (2020). Revisiting the Zintl–Klemm Concept for $ALn_2Ag_3Te_5$-Type Alkaline-Metal (A) Lanthanide (Ln) Silver Tellurides, *Crystals.* **10**, 184.

Einstein, A. (1905). Zur Elektrodynamik bewegter Körper, *Ann. Phys.* **322**, 891–921.

Englert, U. (2017). Single-Crystal X-ray Diffraction, *Handbook of Solid State Chemistry* (edited by Dronskowski, R., Kikkawa, S., & Stein, A.), volume 3, p. 1–28, Wiley-VCH, Weinheim, New York.

Entel, P. (2017). Basic Properties of Well-Known Intermetallics and Some New Complex Magnetic Intermetallics, *Handbook of Solid State Chemistry* (edited by Dronskowski, R., Kikkawa, S., & Stein, A.), volume 5, p. 345–403, Wiley-VCH, Weinheim, New York.

Erba, A., Desmarais, J. K., Casassa, S., Civalleri, B., Donà, L., Bush, I. J., Searle, B., Maschio, L., Edith-Daga, L., Cossard, A., Ribaldone, C., Ascrizzi, E., Marana, N. L., Flament, J.-P., & Kirtman, B. (2022). CRYSTAL23: A Program for Computational Solid State Physics and Chemistry, *J. Chem. Theory Comput.* https://doi.org/10.1021/acs.jctc.2c00958.

Ertural, C., Steinberg, S., & Dronskowski, R. (2019). Development of a Robust Tool to Extract Mulliken and Löwdin Charges from Plane Waves and Its Application to Solid-state Materials, *RSC Adv.* **9**, 29821–29830.

Ertural, C., Stoffel, R. P., Müller, P. C., Vogt, C. A., & Dronskowski, R. (2022). First-Principles Plane-Wave-Based Exploration of Cathode and Anode Materials for Li and Na-ion Batteries Involving Complex Nitrogen-based Anions, *Chem. Mater.* **34**, 652–668.

Esser, M., Deringer, V. L., Wuttig, M., & Dronskowski, R. (2015). Orbital Mixing in Solids as a Descriptor for Materials Mapping, *Solid State Commun.* **203**, 31–34.

Esser, M., Maintz, S., & Dronskowski, R. (2017). Automated First-Principles Mapping for Phase-Change Materials, *J. Comput. Chem.* **38**, 620–628.

Fock, V. (1930a). Näherungsmethode zur Lösung des quantenmechanischen Mehrkörperproblems, *Z. Phys.* **61**, 126–148.

Fock, V. (1930b). "Selfconsistent Field" mit Austausch für Natrium, *Z. Phys.* **62**, 795–805.

Fokwa, B. P. T., Lueken, H., & Dronskowski, R. (2007). Rational Synthetic Tuning Between Itinerant Antiferromagnetism and Ferromagnetism in the Complex Boride Series $Sc_2FeRu_{5-n}Rh_nB_2$ ($0 \leq n \leq 5$), *Chem. Eur. J.* **13**, 6040–6046.

Fredrickson, D. C. (2012). DFT-Chemical Pressure Analysis: Visualizing the Role of Atomic Size in Shaping the Structures of Inorganic Materials, *J. Am. Chem. Soc.* **134**, 5991–5999.

Frenking, G., Loschen, C., Krapp, A., Fau, S., & Strauss, S. H. (2007). Electronic Structure of CO – An Exercise in Modern Chemical Bonding Theory, *J. Comput. Chem.* **28**, 117–126.

Frenking, G. (2022). Heretical Thoughts About the Present Understanding and Description of the Chemical Bond, *Mol. Phys.* e2110168.

Fricke, B. (1975). Superheavy Elements, in *Structure and Bonding* Vol. 21 (edited by Dunitz, J. D.), Springer-Verlag, New York, Heidelberg, Berlin.

Gavezzotti, A. (2008). Non-conventional Bonding Between Organic Molecules. The 'Halogen Bond' in Crystalline Systems, *Mol. Phys.* **106**, 1473–1485.

Gavroglu, K. & Simões, A. (2012). *Neither Physics nor Chemistry. A History of Quantum Chemistry*, The MIT Press, Cambridge, Massachussets, London, England.

George, J., Deringer, V. L., & Dronskowski, R. (2014). Cooperativity of Halogen, Chalcogen, and Pnictogen Bonds in Infinite Molecular Chains by Electronic Structure Theory, *J. Phys. Chem. A.* **118**, 3193–3200.

George, J., Reimann, C., Deringer, V. L., Bredow, T., & Dronskowski, R. (2015). On the DFT Ground State of Crystalline Bromine and Iodine, *ChemPhysChem.* **16**, 728–732.

George, J. & Dronskowski, R. (2017). Tetrel Bonds in Infinite Molecular Chains by Electronic Structure Theory and Their Role for Crystal Stabilization, *J. Phys. Chem. A.* **121**, 1381–1387.

George, J., Waroquiers, D., Di Stefano, D., Petretto, G., Rignanese, G.-M., & Hautier, G. (2020). The Limited Predictive Power of the Pauling Rules, *Angew. Chem. Int. Ed.* **59**, 7569–7575.

Goedecker, S. & Saha, S. (2017). Eliminating Core Electrons in Electronic Structure Calculations: Pseudopotentials and PAW Potentials, *Handbook of Solid State Chemistry* (edited by Dronskowski, R., Kikkawa, S., & Stein, A.), volume 5, p. 29–58, Wiley-VCH, Weinheim, New York.

Görne, A. L., George, J., van Leusen, J., Dück, G., Jacobs, P., Muniraju, N. K. C., & Dronskowski, R. (2016). Ammonothermal Synthesis, Crystal Structure, and Properties of the Ytterbium(II) and Ytterbium(III) Amides and the First Two Rare-Earth Metal Guanidinates, $YbC(NH)_3$ and $Yb(CN_3H_4)_3$, *Inorg. Chem.* **55**, 6161–6168.

Görne, A. L. & Dronskowski, R. (2019). Covalent Bonding versus Total Energy: On the Attainability of Certain Predicted Low-energy Carbon Allotropes, *Carbon.* **148**, 151–158.

Görne, A. L., Scholz, T., Kobertz, D., & Dronskowski, R. (2021). Deprotonating Melamine to Gain Highly Interconnected Materials: Melaminate Salts of Potassium and Rubidium, *Inorg. Chem.* **60**, 15069–15077.

Greenwood, N. N. (1968). *Ionic Crystals, Lattice Defects and Nonstoichiometry*, Butterworth & Co, London.

Greenwood, N. N. & Earnshaw, A. (1997). *Chemistry of the Elements*, 2nd ed., Butterworth-Heinemann, Oxford.

Grimme, S., Ehrlich, S., & Goerigk, L. (2011). Effect of the Damping Function in Dispersion Corrected Density Functional Theory, *J. Comput. Chem.* **32**, 1456–1465.

Grimme, S., Hansen, A., Brandenburg, J. G., & Bannwarth, C. (2016). Dispersion-Corrected Mean-Field Electronic Structure Methods, *Chem. Rev.* **116**, 5105–5154.

Grochala, W., Hoffmann, R., Feng, J., & Ashcroft, N. W. (2007). The Chemical Imagination at Work in *Very Tight Places*, *Angew. Chem. Int. Ed.* **46**, 3620–3642.

Guthrie, F. (1863). XXVIII. – On the Iodide of Iodammonium, *J. Chem. Soc.* **16**, 239–244.

Häussermann, U., Simak, S. I., Ahuja, R., Johansson, B., & Lidin, S. (1999). The Origin of the Distorted Close-Packed Elemental Structure of Indium, *Angew. Chem. Int. Ed.* **38**, 2017–2020.

Hahn, O. & Strassmann, F. (1939). Über den Nachweis und das Verhalten der bei der Bestrahlung des Urans mittels Neutronen entstehenden Erdalkalimetalle, *Naturwissenschaften.* **27**, 11–15.

Hannay, N. B. & Smyth, C. P. (1946). The Dipole Moment of Hydrogen Fluoride and the Ionic Character of Bonds, *J. Am. Chem. Soc.* **68**, 171–173.

Harris, J. & Jones, R. O. (1979). Bonding Trends in the Group-IVA Dimers C_2–Pb_2, *Phys. Rev. A.* **19**, 1813–1818.

Harrison, W. A. (1980). *Electronic Structure and the Properties of Solids: The Physics of the Chemical Bond*, Freeman & Co, San Francisco.

He, L., Jin, Z., Lu, J., & Tang, J. (2002). Modulated Structures of Fe–10Mn–2Cr–1.5C Alloy, *Mater. Des.* **23**, 717–720.

Heitler, W. & London, F. (1927). Wechselwirkung neutraler Atome und homöopolare Bindung nach der Quantenmechanik, *Z. Phys.* **44**, 455–472.

Helgaker, T., Jørgensen, P., & Olsen, J. (2000). *Molecular Electronic-Structure Theory*, John Wiley & Sons Ltd, Chichester.

Hellmann, H. (1935). A New Approximation Method in the Problem of Many Electrons, *J. Chem. Phys.* **3**, 61.

Hellmann, H. (1937a). *Kvantovaya Khimiya*, ONTI, Moscow, Leningrad.

Hellmann, H. (1937b). *Einführung in die Quantenchemie*, Franz Deuticke, Leipzig, Wien.

Hempelmann, J., Müller, P. C., Konze, P. M., Stoffel, R. P., Steinberg, S., & Dronskowski, R. (2021). Long-Range Forces in Rock-Salt-Type Tellurides and How they Mirror the Underlying Chemical Bonding, *Adv. Mater.* **33**, 2100163.

Hempelmann, J., Müller, P. C., Ertural, C., & Dronskowski, R. (2022). The Orbital Origins of Chemical Bonding in Ge–Sb–Te Phase-Change Materials, *Angew. Chem. Int. Ed.* e202115778.

Hoedl, M. F., Ertural, C., Merkle, R., Dronskowski, R., & Maier, J. (2022). The Orbital Nature of Electron Holes in $BaFeO_3$ and Implications for Defect Chemistry, *J. Phys. Chem. C.* **126**, 12809–12819.

Höhn, P. & Niewa, R. (2017). Nitrides of Non-Main Group Elements, *Handbook of Solid State Chemistry* (edited by Dronskowski, R., Kikkawa, S., & Stein, A.), volume 1, p. 251–359, Wiley-VCH, Weinheim, New York.

Hoepfner, V. & Dronskowski, R. (2011). $RbCN_3H_4$: The First Structurally Characterized Salt of a New Class of Guanidinate Compounds, *Inorg. Chem.* **50**, 3799–3803.

Hoepfner, V., Deringer, V. L., & Dronskowski, R. (2012). Hydrogen-Bonding Networks from First Principles: Exploring the Guanidine Crystal, *J. Phys. Chem. A.* **116**, 4551–4559.

Hoepfner, V., Jacobs, P., Sawinski, P. K., Houben, A., Reim, J., & Dronskowski, R. (2013). $RbCN_3H_4$ and $CsCN_3H_4$: A Neutron Powder and Single-Crystal X-ray Diffraction Study, *Z. Anorg. Allg. Chem.* **639**, 1232–1236.

Hoffmann, R. (1963). An Extended Hückel Theory. I. Hydrocarbons, *J. Chem. Phys.* **39**, 1397–1412.

Hoffmann, R. (1988). *Solids and Surfaces: A Chemist's View of Bonding in Extended Structures*, VCH, Weinheim, New York.

Hoffmann, R. (2013). Small but Strong Lessons from Chemistry for Nanoscience, *Angew. Chem. Int. Ed.* **52**, 93–103.

Hoffmann, R., Kabanov, A. A., Golov, A. A., & Proserpio, D. M. (2016). *Homo Citans* and Carbon Allotropes: For an Ethics of Citation, *Angew. Chem. Int. Ed.* **55**, 10962–10976.

Holleman, A. F., Wiberg, E., & Wiberg, N. (2007). *Lehrbuch der Anorganischen Chemie*, 102nd ed., De Gruyter, Berlin.

Hohenberg, P. & Kohn, W. (1964). Inhomogeneous Electron Gas, *Phys. Rev.* **136**, B864–B871.

Hoppe, R., Dähne, W., Mattauch, H., & Rodder, K. M. (1962). Fluorination of Xenon, *Angew. Chem. Int. Ed. Engl.* **1**, 599.

Hoppe, R. (1964). Valence Compounds of the Inert Gases, *Angew. Chem. Int. Ed. Engl.* **3**, 538–545.

Hu, H.-S., Qiu, Y.-H., Xiong, X.-G., Schwarz, W. H. E., & Li, J. (2012). On the Maximum Bond Multiplicity of Carbon: Unusual C≡U Quadruple Bonding in Molecular CUO, *Chem. Sci.* **3**, 2786–2796.

Huang, J.-X., Csányi, G., Zhao, J.-B., Cheng, J., & Deringer, V. L. (2019). First-Principles Study of Alkali-Metal Intercalation in Disordered Carbon Anode Materials, *J. Mater. Chem. A.* **7**, 19070–19080.

Hughbanks, T. & Hoffmann, R. (1983). Chains of Trans-Edge-Sharing Molybdenum Octahedra: Metal–Metal Bonding in Extended Systems, *J. Am. Chem. Soc.* **105**, 3528–3537.

Huppertz, H., Heymann, G., Schwarz, U., & Schwarz, M. R. (2017). High-Pressure Methods in Solid-State Chemistry, *Handbook of Solid State Chemistry* (edited by Dronskowski, R., Kikkawa, S., & Stein, A.), volume 2, p. 23–48, Wiley-VCH, Weinheim, New York.

Hwang, I.-C. & Seppelt, K. (2001). Gold Pentafluoride: Structure and Fluoride Ion Affinity, *Angew. Chem. Int. Ed.* **40**, 3690–3693.

Inkson, J. C. (1986). *Many-Body Theory of Solids – An Introduction*, 2nd print, Plenum Press, New York, London.

Itoh, M. (2017). Structures and Properties of Dielectrics and Ferroelectrics, *Handbook of Solid State Chemistry* (edited by Dronskowski, R., Kikkawa, S., & Stein, A.), volume 1, p. 643–664, Wiley-VCH, Weinheim, New York.

Jeffrey, G. A. (1997). *An Introduction to Hydrogen Bonding*, Oxford University Press, Oxford.

Jensen, W. B. (1996). Electronegativity from Avogadro to Pauling, Part I: Origins of the Electronegativity Concept, *J. Chem. Educ.* **73**, 11–20.

Jayaraman, A., Klement, W., Jr., & Kennedy, G. C. (1963). Phase Diagrams of Calcium and Strontium at High Pressures, *Phys. Rev.* **132**, 1620–1624.

Jones, H. (1934). Applications of the Bloch Theory to the Study of Alloys and of the Properties of Bismuth, *Proc. R. Soc. A.* **147**, 396–417.

Jones, R. O. & Gunnarsson, O. (1989). The Density Functional Formalism, Its Applications and Prospects, *Rev. Mod. Phys.* **61**, 689–746.

Jones, R. O. (2015). Density Functional Theory: Its Origins, Rise to Prominence, and Future, *Rev. Mod. Phys.* **87**, 897–923.

Jones, R. O. (2022). The Chemical Bond in Solids – Revisited, *J. Phys.: Condens. Matter.* **34**, 343001.

Jones, R. O., Elliott, S. R., & Dronskowski, R. (2023). The Myth of "Metavalency" in Phase-Change Materials, *Adv. Mater.* **35**, 2300836.

Jug, K., Ertmer, W., Heidberg, J., Heinemann, H., & Schwarz, W. H. E. (2004). Pionier der Modernen Quantenchemie: Hans Hellmann, *Chem. unserer Zeit.* **37**, 412–421.

Karpov, A., Nuss, J., Wedig, U., & Jansen, M. (2003). Cs$_2$Pt: A Platinide(–II) Exhibiting Complete Charge Separation, *Angew. Chem. Int. Ed.* **42**, 4818–4821.

Kato, D., Tomita, O., Nelson, R., Kirsanova, M. A., Dronskowski, R., Suzuki, H., Zhong, C., Tassel, C., Ishida, K., Matsuzaki, Y., Brown, C. M., Fujita, K., Fujii, K., Yashima, M., Kobayashi, Y., Saeki, A., Oikawa, I., Takamura, H., Abe, R., Kageyama, H., Gorelik, T. E., & Abakumov, A. M. (2022). Bi$_{12}$O$_{17}$Cl$_2$ with a Sextuple Bi–O Layer Composed of Rock-Salt and Fluorite Units and Its Structural Conversion Through Fluorination to Enhance Photocatalytic Activity, *Adv. Funct. Mater.* **32**, 2204112.

Kaupp, M. (2007). The Role of Radial Nodes of Atomic Orbitals for Chemical Bonding and the Periodic Table, *J. Comput. Chem.* **28**, 320–325.

Kauzlarich, S. M. (1996). *Chemistry, Structure and Bonding of Zintl Phases and Ions*, Wiley-VCH, Weinheim, New York.

Ketelaar, J. A. A. (1958). *Chemical Constitution: An Introduction to the Theory of the Chemical Bond*, 2nd rev. ed., Elsevier, Amsterdam.

Koch, W. & Holthausen, M. C. (2001). *A Chemist's Guide to Density Functional Theory*, 2nd ed., Wiley-VCH, Weinheim, New York.

Kohn, W. & Sham, L. J. (1965). Self-Consistent Equations Including Exchange and Correlation Effects, *Phys. Rev.* **140**, A1133–A1138.

Konze, P. M., Dronskowski, R., & Deringer, V. L. (2019). Exploring Chemical Bonding in Phase-Change Materials with Orbital-Based Indicators, *Phys. Stat. Solidi RRL.* **13**, 1800579.

Küpers, M., Konze, P. M., Maintz, S., Steinberg, S., Mio, A. M., Cojocaru-Mirédin, O., Zhu, M., Müller, M., Luysberg, M., Mayer, J., Wuttig, M., & Dronskowski, R. (2017). Unexpected Ge–Ge Contacts in the Two-

Dimensional Ge₄Se₃Te Phase and Analysis of Their Chemical Cause with the Density of Energy (DOE) Function, *Angew. Chem. Int. Ed.* **56**, 10204–10208.

Kutzelnigg, W. (1984). Chemical Bonding in Higher Main Group Elements, *Angew. Chem. Int. Ed. Engl.* **23**, 272–295.

Landrum, G. A., Goldberg, N., & Hoffmann, R. (1997). Bonding in the Trihalides (X_3^-), Mixed Trihalides (X_2Y^-) and Hydrogen Bihalides (X_2H^-). The Connection Between Hypervalent, Electron-Rich Three-Center, Donor–Acceptor and Strong Hydrogen Bonding, *J. Chem. Soc. Dalton Trans.* 3605–3613.

Landrum, G. A. & Dronskowski, R. (1999). Ferromagnetism in Transition Metals: A Chemical Bonding Approach, *Angew. Chem. Int. Ed.* **38**, 1389–1393.

Landrum, G. A. & Dronskowski, R. (2000). The Orbital Origins of Magnetism: From Atoms to Molecules to Ferromagnetic Alloys, *Angew. Chem. Int. Ed.* **39**, 1560–1585.

Laniel, D., Dewaele, A., & Garbarino, G. (2018). High Pressure and High Temperature Synthesis of the Iron Pernitride FeN_2, *Inorg. Chem.* **57**, 6245–6251.

Lee, K., Murray, É. D., Kong, L., Lundqvist, B. I., & Langreth, D. C. (2010). Higher-Accuracy van der Waals Density Functional, *Phys. Rev. B.* **82**, 081101.

Lee, S. & Fredrickson, D. C. (2017). Intermetallic Compounds and Alloy Bonding Theory Derived from Quantum Mechanical One-Electron Models, *Handbook of Solid State Chemistry* (edited by Dronskowski, R., Kikkawa, S., & Stein, A.), volume 1, p. 1–72, Wiley-VCH, Weinheim, New York.

Lee, T. H. & Elliott, S. R. (2021). Multi-center Hyperbonding in Phase-change Materials, *Phys. Stat. Solidi RRL.* **15**, 2000516.

Lenchuk, O., Adelhelm, P., & Mollenhauer, D. (2019). New Insights into the Origin of Unstable Sodium Graphite Intercalation Compounds, *Phys. Chem. Chem. Phys.* **21**, 19378–19390.

Lerch, M., Janek, J., Becker, K. D., Berendts, S., Boysen, H., Bredow, T., Dronskowski, R., Ebbinghaus, S. G., Kilo, M., Lumey, M.-W., Martin, M., Reimann, C., Schweda, E., Valov, I., & Wiemhöfer, H. D. (2009). Oxide Nitrides: From Oxides to Solids with Mobile Nitrogen Ions, *Progr. Solid State Chem.* **37**, 81–131.

Letellier, M., Chevallier, F., Clinard, C., Frackowiak, E., Rouzaud, J.-N., Béguin, F., Morcrette, M., & Tarascon, J.-M. (2003). The First *in situ* ⁷Li Nuclear Magnetic Resonance Study of Lithium Insertion in Hard-Carbon Anode Materials for Li-Ion Batteries, *J. Chem. Phys.* **118**, 6038–6045.

Lewis, G. N. (1916). The Atom and the Molecule, *J. Am. Chem. Soc.* **38**, 762–785.

Li, Y., George, J., Liu, X., & Dronskowski, R. (2015). Synthesis, Structure Determination and Electronic Structure of Magnesium Nitride Chloride, Mg_2NCl, *Z. Anorg. Allg. Chem.* **641**, 266–269.

Li, W.-L., Ertural, C., Bogdanovski, D., Li, J., & Dronskowski, R. (2018a). Chemical Bonding of Crystalline LnB_6 (Ln = La–Lu) and Its Relationship with Ln_2B_8 Gas-Phase Complexes, *Inorg. Chem.* **57**, 12999–13008.

Li, M., Lu, J., Chen, Z., & Amine, K. (2018b). 30 Years of Lithium-Ion Batteries, *Adv. Mater.* **30**, 1800561.

Lima-de-Faria, J., Hellner, E., Liebau, F., Makovicky, E., & Parthé, E. (1990). Nomenclature of Inorganic Structure Types, *Acta Cryst. A.* **46**, 1–11.

Lintzen, S., von Appen, J., Hallstedt, B., & Dronskowski, R. (2013). The Fe–Mn Enthalpy Phase Diagram from First Principles, *J. Alloys Compd.* **577**, 370–375.

Lipscomb, W. N. (1966). Framework Rearrangement in Boranes and Carboranes, *Science.* **153**, 373–378.

Lipscomb, W. N. (1977). The Boranes and Their Relatives, *Science.* **196**, 1047–1055.

Liu, X., Decker, A., Schmitz, D., & Dronskowski, R. (2000). Crystal Structure Refinement of Lead Cyanamide and the Stiffness of the Cyanamide Anion, *Z. Anorg. Allg. Chem.* **626**, 103–105.

Liu, X., Müller, P., Kroll, P., Dronskowski, R., Wilsmann, W., & Conradt, R. (2003). Experimental and Quantum-Chemical Studies on the Thermochemical Stabilities of Mercury Carbodiimide and Mercury Cyanamide, *ChemPhysChem.* **4**, 725–731.

Liu, X., Krott, M., Müller, P., Hu, C., Lueken, H., & Dronskowski, R. (2005). Synthesis, Crystal Structure, and Properties of MnNCN, the First Carbodiimide of a Magnetic Transition Metal, *Inorg. Chem.* **44**, 3001–3003.

Liu, X., Stork, L., Speldrich, M., Lueken, H., & Dronskowski, R. (2009). FeNCN and Fe(NCNH)$_2$: Synthesis, Structure and Magnetic Properties of a Nitrogen-Based Pseudo-Oxide and -Hydroxide of Divalent Iron, *Chem. Eur. J.* **15**, 1558–1561.

Liu, X., Wessel, C., Pan, F., & Dronskowski, R. (2013). Synthesis and Single-Crystal Structure Determination of the Zinc Nitride Halides Zn$_2$NX (X = Cl, Br, I), *J. Solid State Chem.* **203**, 31–36.

Liu, X., George, J., Maintz, S., & Dronskowski, R. (2015). β-CuN$_3$: The Overlooked Ground-State Polymorph of Copper Azide with Heterographene-Like Layers, *Angew. Chem. Int. Ed.* **54**, 1954–1959.

Liu, X., Stoffel, R., & Dronskowski, R. (2020). Syntheses, Crystal Structures, and Vibrational Properties of Two Lead Azide Halides PbN$_3$X (X = Cl, Br), *Z. Anorg. Allg. Chem.* **646**, 1525–1530.

London, F. (1930). Zur Theorie und Systematik der Molekularkräfte, *Z. Phys.* **63**, 245–279.

Longuet-Higgins, H. G. (1949). Substances hydrogénées avec défaut d'électrons, *J. Chim. Phys.* **46**, 268–275.

Lonie, D. C. & Zurek, E. (2011). Xtalopt: An Open-Source Evolutionary Algorithm for Crystal Structure Prediction, *Comput. Phys. Commun.* **182**, 372–387.

Lüdtke, T., Schmidt, A., Göbel, C., Fischer, A., Becker, N., Reimann, C., Bredow, T., Dronskowski, R., & Lerch, M. (2014). Synthesis and Crystal Structure of δ-TaON, a Metastable Polymorph of Tantalum Oxide Nitride, *Inorg. Chem.* **53**, 11691–11698.

Lüdtke, T., Weber, D., Schmidt, A., Müller, A., Reimann, C., Becker, N., Bredow, T., Dronskowski, R., Ressler, T., & Lerch, M. (2017). Synthesis and Characterization of Metastable Transition Metal Oxides and Oxide Nitrides, *Z. Kristallogr. Cryst. Mater.* **232**, 3–14.

Lumey, M.-W. & Dronskowski, R. (2003). The Electronic Structure of Tantalum Oxynitride and the Falsification of α-TaON, *Z. Anorg. Allg. Chem.* **629**, 2173–2179.

Lumey, M.-W. & Dronskowski, R. (2005). First-Principles Electronic Structure, Chemical Bonding and High-Pressure Phase Prediction of the Oxynitrides of Vanadium, Niobium and Tantalum, *Z. Anorg. Allg. Chem.* **631**, 887–893.

Luo, D., Qiao, X., & Dronskowski, R. (2021). Predicting Nitrogen-Based Families of Compounds: Transition-Metal Guanidinates TCN$_3$ (T = V, Nb, Ta) and Ortho-Nitrido Carbonates T'_2CN$_4$ (T' = Ti, Zr, Hf), *Angew. Chem. Int. Ed.* **60**, 486–492.

Luo, D., Yin, K., & Dronskowski, R. (2022). Existence of BeCN$_2$ and Its First-Principles Phase Diagram: Be and C Introducing Structural Diversity, *J. Am. Chem. Soc.* **144**, 5155–5162.

Ma, Y., Eremets, M., Oganov, A. R., Xie, Y., Trojan, I., Medvedev, S., Lyakhov, A. O., Valle, M., & Prakapenka, V. (2009). Transparent Dense Sodium, *Nature.* **458**, 182–185.

Maier, J. (2017). Defect Chemistry and Its Relevance for Ionic Conduction and Reactivity, *Handbook of Solid State Chemistry* (edited by Dronskowski, R., Kikkawa, S., & Stein, A.), volume 1, p. 665–701, Wiley-VCH, Weinheim, New York.

Maintz, S., Deringer, V. L., Tchougréeff, A. L., & Dronskowski, R. (2013). Analytic Projection from Plane-Wave and PAW Wavefunctions and Application to Chemical-Bonding Analysis in Solids, *J. Comput. Chem.* **34**, 2557–2567.

Maintz, S., Deringer, V. L., Tchougréeff, A. L., & Dronskowski, R. (2016). LOBSTER: A Tool to Extract Chemical Bonding from Plane-Wave Based DFT, *J. Comput. Chem.* **37**, 1030–1035.

Matthias, B. T. (1955). Empirical Relation Between Superconductivity and the Number of Valence Electrons per Atom, *Phys. Rev.* **97**, 74–76.

Mayer, I. (1983). Charge, Bond Order and Valence in the Ab Initio SCF Theory, *Chem. Phys. Lett.* **97**, 270–274.

Mayer, I. (2003). *Simple Theorems, Proofs, and Derivations in Quantum Chemistry*, Kluwer Academic/Plenum Publishers, New York.

Mayer, I. (2017). *Bond Orders and Energy Components*, CRC Press, Boca Raton, London, New York.

McCalla, E. (2017). Electrical Energy Storage: Batteries, *Handbook of Solid State Chemistry* (edited by Dronskowski, R., Kikkawa, S., & Stein, A.), volume 6, p. 1–24, Wiley-VCH, Weinheim, New York.

Meissner, E. & Niewa, R. (Eds.) (2021). *Ammonothermal Synthesis and Crystal Growth of Nitrides*, Springer Nature Switzerland AG.

Meister, J. & Schwarz, W. H. E. (1994). Principal Components of Ionicity, *J. Phys. Chem.* **98**, 8245–8252.

Mendelejeff, D. (1869). Über die Beziehungen der Eigenschaften zu den Atomgewichten der Elemente, *Z. Chem.* **12**, 405–406.

Meven, M. & Roth, G. (2017). Neutron Diffraction, *Handbook of Solid State Chemistry* (edited by Dronskowski, R., Kikkawa, S., & Stein, A.), volume 3, p. 77–108, Wiley-VCH, Weinheim, New York.

Meyer, L. (1864). *Die Modernen Theorien der Chemie*, Maruschke & Berendt, Breslau.

Miller, G. J., Zhang, Y., & Wagner, F. R. (2017). Chemical Bonding in Solids, *Handbook of Solid State Chemistry* (edited by Dronskowski, R., Kikkawa, S., & Stein, A.), volume 5, p. 405–489, Wiley-VCH, Weinheim, New York.

Missong, R., George, J., Houben, A., & Dronskowski, R. (2015). Synthesis, Structure, and Properties of SrC(NH)$_3$, a Nitrogen-Based Carbonate Analogue with the Trinacria Motif, *Angew. Chem. Int. Ed.* **54**, 12171–12175.

Monacelli, L., Casula, M., Nakano, K., Sorella, S., & Mauri, F. (2023). Quantum Phase Diagram of High-Pressure Hydrogen, *Nature Phys.* **19**, 845–850.

Monkhorst, H. J. & Pack, J. D. (1976). Special Points for Brillouin-Zone Integrations, *Phys. Rev. B.* **13**, 5188–5192.

Moore, W. J. (1967). *Seven Solid States*, W. A. Benjamin Inc, New York.

Müller, B. G. (1987). Fluorides of Copper, Silver, Gold, and Palladium, *Angew. Chem. Int. Ed. Engl.* **26**, 1081–1097.

Müller, U. (2006). *Inorganic Structural Chemistry*, 2nd ed., John Wiley & Sons, Hoboken, New Jersey.

Müller, P. C., Ertural, C., Hempelmann, J., & Dronskowski, R. (2021). Crystal Orbital Bond Index: Covalent Bond Orders in Solids, *J. Phys. Chem. C.* **125**, 7959–7970.

Mulliken, R. S. (1955a). Electronic Population Analysis on LCAO-MO Molecular Wave Functions. I, *J. Chem. Phys.* **23**, 1833–1840.

Mulliken, R. S. (1955b). Electronic Population Analysis on LCAO-MO Molecular Wave Functions. II. Overlap Populations, Bond Orders, and Covalent Bond Energies, *J. Chem. Phys.* **23**, 1841–1846.

Munzarová, M. L. & Hoffmann, R. (2002). Electron-Rich Three-Center Bonding: Role of s, p interactions Across the p-block, *J. Am. Chem. Soc.* **124**, 4787–4795.

Nakada, A., Kato, D., Nelson, R., Takahira, H., Yabuuchi, M., Higashi, M., Suzuki, H., Kirsanova, M., Kakudou, N., Tassel, C., Yamamoto, T., Brown, C. M., Dronskowski, R., Saeki, A., Abakumov, A., Kageyama, H., & Abe, R. (2021). Conduction Band Control of Oxyhalides with a Triple-Fluorite Layer for Visible Light Photocatalysis, *J. Am. Chem. Soc.* **143**, 2491–2499.

Nagamatsu, J., Nakagawa, N., Muranaka, T., Zenitani, Y., & Akimitsu, J. (2001). Superconductivity at 39 K in Magnesium Diboride, *Nature.* **410**, 63–64.

Nandi, A. & Kozuch, S. (2020). History and Future of Dative Bonds, *Chem. Eur. J.* **26**, 759–772.

Nelson, R., Konze, P. M., & Dronskowski, R. (2017). First-Principles Chemical Bonding Study of Manganese Carbodiimide, MnNCN, as Compared to Manganese Oxide, MnO, *J. Phys. Chem. A.* **121**, 7778–7786.

Nelson, R., Ertural, C., George, J., Deringer, V. L., Hautier, G., & Dronskowski, R. (2020). LOBSTER: Local Orbital Projections, Atomic Charges, and Chemical-Bonding Analysis from Projector-Augmented-Wave-Based Density-Functional Theory, *J. Comput. Chem.* **41**, 1931–1940.

Nelson, R., Ertural, C., Müller, P. C., & Dronskowski, R. (2023). Chemical Bonding with Plane Waves, in *Comprehensive Inorganic Chemistry III* Vol. 3 (edited by Reedijk, J. & Poeppelmeier, K.), p. 141–201, Elsevier.

Niewa, R. & Jacobs, H. (1996). Group V and VI Alkali Nitridometalates: A Growing Class of Compounds with Structures Related to Silicate Chemistry, *Chem. Rev.* **96**, 2053–2062.

Niewa, R. & DiSalvo, F. J. (1998). Recent Developments in Nitride Chemistry, *Chem. Mater.* **10**, 2733–2752.

Oganov, A. R. (Ed.) (2010). *Modern Methods of Crystal Structure Prediction*, Wiley-VCH, Weinheim, New York.

Ogg, R. A., Jr. (1946). Bose–Einstein Condensation of Trapped Electron Pairs. Phase Separation and Superconductivity of Metal-Ammonia Solutions, *Phys. Rev.* **69**, 243–244.

Otero de la Roza, A. & DiLabio, G. A. (2017). *Non-Covalent Interactions in Quantum Chemistry and Physics: Theory and Applications*, Elsevier, Amsterdam, Netherlands.

Papoian, G. A. & Hoffmann, R. (2000). Hypervalent Bonding in One, Two, and Three Dimensions: Extending the Zintl–Klemm Concept to Nonclassical Electron-Rich Networks, *Angew. Chem. Int. Ed.* **39**, 2408–2448.

Pauli, W. (1925). Über den Zusammenhang des Abschlusses der Elektronengruppen im Atom mit der Komplexstruktur der Spektren, *Z. Phys.* **31**, 765–783.

Pauling, L. (1960). *The Nature of the Chemical Bond*, Cornell University Press, Ithaca, New York.

Parr, R. G. & Pearson, R. G. (1983). Absolute Hardness: Companion Parameter to Absolute Electronegativity, *J. Am. Chem. Soc.* **105**, 7512–7516.

Parr, R. G. & Yang, W. (1989). *Density Functional Theory of Atoms and Molecules*, Clarendon Press, New York.

Pearson, R. G. (1988). Absolute Electronegativity and Hardness: Application to Inorganic Chemistry, *Inorg. Chem.* **27**, 734–740.

Pearson, R. G. (1997). *Chemical Hardness*, Wiley-VCH, Weinheim, New York.

Peierls, R. E. (1955). *Quantum Theory of Solids*, Oxford University Press, Oxford.

Peierls, R. E. (1993). Early Work on Solids, Mainly in the Thirties, *Rev. Mod. Phys.* **65**, 251–254.

Pimentel, G. C. (1951). The Bonding of Trihalide and Bifluoride Ions by the Molecular Orbital Method, *J. Chem. Phys.* **19**, 446–448.

Pimentel, G. C. & McClellan, A. L. (1960). *The Hydrogen Bond*, W. H. Freeman, San Francisco.

Plekhanov, E. A. & Tchougréeff, A. L. (2017). Resonating Valence Bonds in Chemistry and Solid State, *Handbook of Solid State Chemistry* (edited by Dronskowski, R., Kikkawa, S., & Stein, A.), volume 5, p. 87–117, Wiley-VCH, Weinheim, New York.

Plekhanov, E. A., Tchougréeff, A. L., & Dronskowski, R. (2020). ΘΦ: Solid State Package Allowing Bardeen–Cooper–Schrieffer and Magnetic Superstructure Electronic States, *Comput. Phys. Commun.* **251**, 107079.

Popov, I. V., Görne, A. L., Tchougréeff, A. L., & Dronskowski, R. (2019). Relative Stability of Diamond and Graphite as Seen Through Bonds and Hybridizations, *Phys. Chem. Chem. Phys.* **21**, 10961–10969.

Pöttgen, R. & Johrendt, D. (2019). *Intermetallics: Synthesis, Structure, Function*, 2nd Ed., De Gruyter, Berlin, Boston.

Pöttgen, R., Corkett, A. J., & Dronskowski, R. (2023). NiAs-Derived Cyanamide (Carbodiimide) Structures – A Group-Theoretical View, *Z. Kristallogr.* **238**, 95–103.

Primas, H. (1981). *Chemistry, Quantum Mechanics and Reductionism*, Springer-Verlag, Berlin, Heidelberg.

Primas, H. & Müller-Herold, U. (1984). *Elementare Quantenchemie*, Teubner, Stuttgart, Leipzig.

Pyykkö, P. (1988). Relativistic Effects in Structural Chemistry, *Chem. Rev.* **88**, 563–594.

Qiao, X., Liu, X., Bayarjargal, L., Corkett, A. J., Wang, W., Ma, Z., Lin, Z., & Dronskowski, R. (2021). Nonlinear Optical Effects in Two Mercury Cyanamide/Guanidinium Chlorides $Hg_3(NCN)_2Cl_2$ and $Hg_2(C(NH_2)_3)Cl_5$, *J. Mater. Chem. C.* **9**, 967–974.

Rahm, M., Zeng, T., & Hoffmann, R. (2019). Electronegativity Seen as the Ground-State Average Valence Electron Binding Energy, *J. Am. Chem. Soc.* **141**, 342–351.

Rao, F., Ding, K., Zhou, Y., Zheng, Y., Xia, M., Lv, S., Song, Z., Feng, S., Ronneberger, I., Mazzarello, R., Zhang, W., & Ma, E. (2017). Reducing the Stochasticity of Crystal Nucleation to Enable Subnanosecond Memory Writing, *Science.* **358**, 1423–1427.

Reddy, R. R., Gopal, K. R., Ahammed, Y. N., Narasimhulu, K., Reddy, L. S. S., & Reddy, C. V. K. (2005). Correlation between Optical Electronegativity, Molar Refraction, Ionicity and Density of Binary Oxides, Silicates and Minerals, *Solid State Ion.* **176**, 401–407.

Reitz, L. S., Müller, P. C., Schnieders, D., Dronskowski, R., Choi, W. I., Son, W.-J., Jang, I., & Kim, D. S. (2023). On the Atomistic Origin of the Polymorphism and the Dielectric Physical Properties of Beryllium Oxide, *J. Comput. Chem.* **44**, 1052–1063.

Riedel, R., Wiehl, L., Zerr, A., Zinin, P., & Kroll, P. (2017). Superhard Materials, *Handbook of Solid State Chemistry* (edited by Dronskowski, R., Kikkawa, S., & Stein, A.), volume 6, p. 175–200, Wiley-VCH, Weinheim, New York.

Rotter, M., Tegel, M., & Johrendt, D. (2008). Superconductivity at 38 K in the Iron Arsenide (Ba$_{1-x}$K$_x$)Fe$_2$As$_2$, *Phys. Rev. Lett.* **101**, 107006.

Saito, G. & Yoshida, Y. (2011). Organic Superconductors, *Chem. Rec.* **11**, 124–145.

Sánchez-Portal, D., Artacho, E., & Soler, J. M. (1995). Projection of Plane-wave Calculations into Atomic Orbitals, *Solid State Commun.* **95**, 685–690.

Sánchez-Portal, D., Artacho, E., & Soler, J. M. (1996). Analysis of Atomic-Orbital Basis Sets from the Projection of Plane-Wave Results, *J. Phys.: Condens. Matter.* **8**, 3859–3880.

van Santen, R. A. (1984). The Ostwald Step Rule, *J. Phys. Chem.* **88**, 5768–5769.

Sawinski, P. K. & Dronskowski, R. (2012). Solvothermal Synthesis, Crystal Growth, and Structure Determination of Sodium and Potassium Guanidinate, *Inorg. Chem.* **51**, 7425–7430.

Sawinski, P. K., Meven, M., Englert, U., & Dronskowski, R. (2013). Single-Crystal Neutron Diffraction Study on Guanidine, CN$_3$H$_5$, *Cryst. Growth Des.* **13**, 1730–1735.

Schädler, H.-D., Jäger, L., & Senf, I. (1993). Pseudoelementverbindungen. V. Pseudochalkogene – Versuch der empirischen und theoretischen Charakterisierung eines Konzeptes, *Z. Anorg. Allg. Chem.* **619**, 1115–1120.

Schäfer, H., Eisenmann, B., & Müller, W. (1973). Zintl Phases: Transitions between Metallic and Ionic Bonding, *Angew. Chem. Int. Ed. Engl.* **12**, 694–712.

Schilling, H., Stork, A., Irran, E., Wolff, H., Bredow, T., Dronskowski, R., & Lerch, M. (2007). γ-TaON: A Metastable Polymorph of Tantalum Oxynitride, *Angew. Chem. Int. Ed.* **46**, 2931–2934.

Schneider, S. B., Seibald, M., Deringer, V. L., Stoffel, R. P., Frankovsky, R., Friederichs, G. M., Laqua, H., Duppel, V., Jeschke, G., Dronskowski, R., & Schnick, W. (2013). High-Pressure Synthesis and Characterization of Li$_2$Ca$_3$[N$_2$]$_3$ – An Uncommon Metallic Diazenide with [N$_2$]$^{2-}$ Ions, *J. Am. Chem. Soc.* **135**, 16668–16679.

von Schnering, H. G. (1981). Homoatomic Bonding of Main Group Elements, *Angew. Chem. Int. Ed. Engl.* **20**, 33–51.

Schrödinger, E. (1926a). Quantisierung als Eigenwertproblem, *Ann. Phys.* **385**, 437–490.

Schrödinger, E. (1926b). Quantisierung als Eigenwertproblem, *Ann. Phys.* **386**, 109–139.

Schummer, J. (1998). The Chemical Core of Chemistry I: A Conceptual Approach, *Hyle.* **4**, 129–162.

Schwarz, K., Blaha, P., & Madsen, G. K. H. (2002). Electronic Structure Calculations of Solids using the WIEN2k Package for Material Sciences, *Comp. Phys. Comm.* **147**, 71–76.

Schwarz, W. H. E. (2006). Measuring Orbitals: Provocation or Reality?, *Angew. Chem. Int. Ed.* **45**, 1508–1517.

Schwarz, U., Wosylus, A., Böhme, B., Baitinger, M., Hanfland, M., & Grin, Y. (2008). A 3D Network of Four-Bonded Germanium: A Link Between Open and Dense, *Angew. Chem. Int. Ed.* **47**, 6790–6793.

Schwarz, W. H. E. (2010). An Introduction to Relativistic Quantum Chemistry, in *Relativistic Methods for Chemists* (edited by Barysz, M. & Ishikawa, Y.), p. 1–62, Springer Science+Business Media B.V.

Schwarz, K. & Blaha, P. (2017). DFT Calculations for Real Solids, *Handbook of Solid State Chemistry* (edited by Dronskowski, R., Kikkawa, S., & Stein, A.), volume 5, p. 237–259, Wiley-VCH, Weinheim, New York.

Scotti, N., Kockelmann, W., Senker, J., Traßel, S., & Jacobs, H. (1999). Sn$_3$N$_4$, ein Zinn(IV)-nitrid – Synthese und erste Strukturbestimmung einer binären Zinn–Stickstoff-Verbindung, *Z. Anorg. Allg. Chem.* **625**, 1435–1439.

Seeman, J. I. (2022a). Why Hoffmann? The Person and the Young Chemical Physicist, *Chem. Rec.* **22**, e202200196.

Seeman, J. I. (2022b). Why Hoffmann? His Chemistry, *Chem. Rec.* **22**, e202200205.

Seifert, G. (2017). Tight-Binding Density Functional Theory: DFTB, *Handbook of Solid State Chemistry* (edited by Dronskowski, R., Kikkawa, S., & Stein, A.), volume 5, p. 203–226, Wiley-VCH, Weinheim, New York.

Seo, D.-K. (2017). Spin Polarization, *Handbook of Solid State Chemistry* (edited by Dronskowski, R., Kikkawa, S., & Stein, A.), volume 5, p. 261–283, Wiley-VCH, Weinheim, New York.

Shen, J., Jia, S., Shi, N., Ge, Q., Gotoh, T., Lv, S., Liu, Q., Dronskowski, R., Elliott, S. R., Song, Z., & Zhu, M. (2021). Elemental Electrical Switch Enabling Phase Segregation-Free Operation, *Science*. **374**, 1390–1394.

Shimakawa, Y. (2017). Perovskite Structure Compounds, *Handbook of Solid State Chemistry* (edited by Dronskowski, R., Kikkawa, S., & Stein, A.), volume 1, p. 221–250, Wiley-VCH, Weinheim, New York.

Simon, A. (1983). Intermetallic Compounds and the Use of Atomic Radii in Their Description, *Angew. Chem. Int. Ed. Engl.* **22**, 95–113.

Simon, A. (1997). Superconductivity and Chemistry, *Angew. Chem. Int. Ed. Engl.* **36**, 1788–1806.

Simon, A. (2015). Superconductivity and the Periodic Table: From Elements to Materials, *Phil. Trans. R. Soc. A*. **373**, 20140192.

Simons, J., Hempelmann, J., Fries, K. S., Müller, P. C., Dronskowski, R., & Steinberg, S. (2021). Bonding Diversity in Rock Salt-Type Tellurides: Examining the Interdependence between Chemical Bonding and Materials Properties, *RSC Adv.* **11**, 20679–20686.

Slater, J. C. (1951). A Simplification of the Hartree–Fock Method, *Phys. Rev.* **81**, 385–390.

Song, W., Zhang, W., von Appen, J., Dronskowski, R., & Bleck, W. (2015). κ-Phase Formation in Fe-Mn-Al-C Austenitic Steels, *Steel Res. Int.* **86**, 1161–1169.

Song, W., Bogdanovski, D., Yildiz, A. B., Houston, J. E., Dronskowski, R., & Bleck, W. (2018). On the Mn–C Short-Range Ordering in a High-Strength High-Ductility Steel: Small Angle Neutron Scattering and Ab Initio Investigation, *Metals*. **8**, 44.

Sougrati, M. T., Darwiche, A., Liu, X., Mahmoud, A., Hermann, R. P., Jouen, S., Monconduit, L., Dronskowski, R., & Stievano, L. (2016). Transition-Metal Carbodiimides as Molecular Negative Electrode Materials for Lithium- and Sodium-Ion Batteries with Excellent Cycling Performances, *Angew. Chem. Int. Ed.* **55**, 5090–5095.

Springborg, M. & Dong, Y. (2017). Density Functional Theory, *Handbook of Solid State Chemistry* (edited by Dronskowski, R., Kikkawa, S., & Stein, A.), volume 5, p. 1–28, Wiley-VCH, Weinheim, New York.

Steinberg, S. & Dronskowski, R. (2018). The Crystal Orbital Hamilton Population (COHP) Method as a Tool to Visualize and Analyze Chemical Bonding in Intermetallic Compounds, *Crystals*. **8**, 225.

Steiner, T. (2002). The Hydrogen Bond in the Solid State, *Angew. Chem. Int. Ed.* **41**, 48–76.

Stoffel, R. P. & Dronskowski, R. (2012). Barium Peroxide: A Simple Test Case for First-Principles Investigations on the Temperature Dependence of Solid-State Vibrational Frequencies, *Z. Anorg. Allg. Chem.* **638**, 1403–1406.

Stoffel, R. P., Deringer, V. L., Simon, R. E., Hermann, R. P., & Dronskowski, R. (2015). A Density-Functional Study on the Electronic and Vibrational Properties of Layered Antimony Telluride, *J. Phys.: Condens. Matter*. **27**, 085402.

Stoffel, R. P. & Dronskowski, R. (2017). Lattice Dynamics and Thermochemistry of Solid-State Materials from First-Principles Quantum-Chemical Calculations, *Handbook of Solid State Chemistry* (edited by Dronskowski, R., Kikkawa, S., & Stein, A.), volume 5, p. 491–526, Wiley-VCH, Weinheim, New York.

Sugden, S. (1930). *The Parachor and Valency*, G. Routledge, London.

Szabo, A. & Ostlund, N. S. (1989). *Modern Quantum Chemistry: Introduction to Advanced Electronic Structure Theory*, McGraw-Hill, New York.

Tang, X., Xiang, H., Liu, X., Speldrich, M., & Dronskowski, R. (2010). A Ferromagnetic Carbodiimide: $Cr_2(NCN)_3$, *Angew. Chem. Int. Ed.* **49**, 4738–4742.

Tang, F., Bogdanovski, D., Bajenova, I., Khvan, A., Dronskowski, R., & Hallstedt, B. (2018). A CALPHAD Assessment of the Al–Mn–C System Supported by *ab initio* Calculations, *CALPHAD*. **60**, 231–239.

Tchougréeff, A. L., Liu, X., Müller, P., van Beek, W., Ruschewitz, U., & Dronskowski, R. (2012). Structural Study of CuNCN and Its Theoretical Implications: A Case of a Resonating-Valence-Bond State?, *J. Phys. Chem. Lett.* **3**, 3360–3366.

Tchougréeff, A. L., Plekhanov, E., & Dronskowski, R. (2021). Solid-State Quantum Chemistry with ΘΦ (*ThetaPhi*): Spin-Liquids, Superconductors, and Magnetic Superstructures Made Computationally Available, *J. Comput. Chem.* **42**, 1498–1513.

Teale, A. M., Helgaker, T., Savin, A., Adamo, C., Aradi, B., Arbuznikov, A. V., Ayers, P. W., Baerends, E. J., Barone, V., Calaminici, P., Cancès, E., Carter, E. A., Chattaraj, P. K., Chermette, H., Ciofini, I., Crawford, T. D., De Proft, F., Dobson, J. F., Draxl, C., Frauenheim, T., Fromager, E., Fuentealba, P., Gagliardi, L., Galli, G., Gao, J., Geerlings, P., Gidopoulos, N., Gill, P. M. W., Gori-Giorgi, P., Görling, A., Gould, T., Grimme, S., Gritsenko, O., Jensen, H. J. A., Johnson, E. R., Jones, R. O., Kaupp, M., Köster, A. M., Kronik, L., Krylov, A. I., Kvaal, S., Laestadius, A., Levy, M., Lewin, M., Liu, S., Loos, P.-F., Maitra, N. T., Neese, F., Perdew, J. P., Pernal, K., Pernot, P., Piecuch, P., Rebolini, E., Reining, L., Romaniello, P., Ruzsinszky, A., Salahub, D. R., Scheffler, M., Schwerdtfeger, P., Staroverov, V. N., Sun, J., Tellgren, E., Tozer, D. J., Trickey, S. B., Ullrich, C. A., Vela, A., Vignale, G., Wesolowski, T. A., Xu, X., & Yang, W. (2022). DFT Exchange: Sharing Perspectives on the Workhorse of Quantum Chemistry and Materials Science, *Phys. Chem. Chem. Phys.* **24**, 28700–28781.

Tessier, F. (2017). Fluorite-Type Transition Metal Oxynitrides, *Handbook of Solid State Chemistry* (edited by Dronskowski, R., Kikkawa, S., & Stein, A.), volume 1, p. 362–382, Wiley-VCH, Weinheim, New York.

Timmerscheidt, T. A., Dey, P., Bogdanovski, D., von Appen, J., Hickel, T., Neugebauer, J., & Dronskowski, R. (2017). The Role of κ-Carbides as Hydrogen Traps in High-Mn Steels, *Metals*. **7**, 264.

Tkatchenko, A. & Scheffler, M. (2009). Accurate Molecular Van Der Waals Interactions from Ground-State Electron Density and Free-Atom Reference Data, *Phys. Rev. Lett.* **102**, 073005.

Tokmachev, A. M., Tchougréeff, A. L., & Dronskowski, R. (2010). Hydrogen-Bond Networks in Water Clusters $(H_2O)_{20}$: An Exhaustive Quantum-Chemical Analysis, *ChemPhysChem*. **11**, 384–388.

Tschauner, O., Luo, S. N., Chen, Y. J., McDowell, A., Knight, J., & Clark, S. M. (2013). Shock Synthesis of Lanthanum-III-Pernitride, *High Press. Res.* **33**, 202–207.

Usvyat, S., Maschio, L., & Schütz, M. (2017). Periodic Møller–Plesset Perturbation Theory of Second Order for Solids, *Handbook of Solid State Chemistry* (edited by Dronskowski, R., Kikkawa, S., & Stein, A.), volume 5, p. 59–86, Wiley-VCH, Weinheim, New York.

Vannerberg, N.-G. (1962). The Crystal Structure of Calcium Cyanamide, *Acta Chem. Scand.* **16**, 2263–2266.

Wang, Y., Lv, J., Zhu, L., & Ma, Y. (2010). Crystal Structure Prediction via Particle-Swarm Optimization, *Phys. Rev. B.* **82**, 094116.

Wang, R., George, J., Potts, S. K., Kremer, M., Dronskowski, R., & Englert, U. (2019). The Many Flavours of Halogen Bonds – Message from Experimental Electron Density and Raman Spectroscopy, *Acta Cryst. C.* **75**, 1190–1201.

Waser, R. & Hoffmann-Eifert, S. (2017). Dielectric Properties, *Handbook of Solid State Chemistry* (edited by Dronskowski, R., Kikkawa, S., & Stein, A.), volume 3, p. 523–560, Wiley-VCH, Weinheim, New York.

Wessel, M. & Dronskowski, R. (2010). Nature of N–N Bonding within High-Pressure Noble-Metal Pernitrides and the Prediction of Lanthanum Pernitride, *J. Am. Chem. Soc.* **132**, 2421–2429.

Wessel, M. & Dronskowski, R. (2011). A New Phase in the Binary Iron Nitrogen System? – The Prediction of Iron Pernitride, FeN_2, *Chem. Eur. J.* **17**, 2598–2603.

Wiberg, K. B. (1968). Application of the Pople-Santry-Segal CNDO Method to the Cyclopropylcarbinyl and Cyclobutyl Cation and to Bicyclobutane, *Tetrahedron*. **24**, 1083–1096.

Wolff, H., Schilling, H., Lerch, M., & Dronskowski, R. (2006). A Density-Functional and Molecular-Dynamics Study on the Physical Properties of Yttrium-Doped Tantalum Oxynitride, *J. Solid State Chem.* **179**, 2265–2270.

Wolff, H., Bredow, T., Lerch, M., Schilling, H., Irran, E., Stork, A., & Dronskowski, R. (2007). A First-Principles Study of the Electronic and Structural Properties of γ-TaON, *J. Phys. Chem. A.* **111**, 2745–2749.

Wolff, H. & Dronskowski, R. (2008). First-Principles and Molecular-Dynamics Study of Structure and Bonding in Perovskite-Type Oxynitrides ABO_2N (A = Ca, Sr, Ba; B = Ta, Nb), *J. Comput. Chem.* **29**, 2260–2267.

Woodhead, K., Pascarelli, S., Hector, A. L., Briggs, R., Alderman, N., & McMillan, P. F. (2014). High Pressure Polymorphism of β-TaON, *Dalton Trans.* **43**, 9647–9654.

Wuttig, M., Lüsebrink, D., Wamwangi, D., Wełnic, W., Gilleßen, M., & Dronskowski, R. (2007). The Role of Vacancies and Local Distortions in the Design of New Phase-change Materials, *Nat. Mater.* **6**, 122–128.

Xi, L., Pan, S., Li, X., Xu, Y., Ni, J., Sun, X., Yang, J., Luo, J., Xi, J., Zhu, W., Li, X., Jiang, D., Dronskowski, R., Shi, X., Snyder, G. J., & Zhang, W. (2018). Discovery of High-Performance Thermoelectric Chalcogenides Through Reliable High-Throughput Material Screening, *J. Am. Chem. Soc.* **140**, 10785–10793.

Yamada, T., Liu, X., Englert, U., Yamane, H., & Dronskowski, R. (2009). Solid-State Structure of Free Base Guanidine Achieved at Last, *Chem. Eur. J.* **15**, 5651–5655.

Yamada, O. (2012). Origin, Secret, and Application of the Ideal Phase-Change Material GeSbTe, *Phys. Status Solidi B.* **249**, 1837–1842.

Yu, X., Oganov, A. R., Wang, Z., Saleh, G., Sharma, V., Zhu, Q., Wang, Q., Zhou, X.-F., Popov, I. A., Boldyrev, A. I., Baturin, V. S., & Lepeshkin, S. V. (2017). Predicting the Structure and Chemistry of Low-Dimensional Materials, *Handbook of Solid State Chemistry* (edited by Dronskowski, R., Kikkawa, S., & Stein, A.), volume 5, p. 527–570, Wiley-VCH, Weinheim, New York.

Zhang, Y., Verbraeken, M. C., Tassel, C., & Kageyama, H. (2017). Metal Hydrides, *Handbook of Solid State Chemistry* (edited by Dronskowski, R., Kikkawa, S., & Stein, A.), volume 1, p. 477–520, Wiley-VCH, Weinheim, New York.

Zucker, U. & Schulz, H. (1982). Statistical Approaches for the Treatment of Anharmonic Motion in Crystals. II. Anharmonic Thermal Vibrations and Effective Atomic Potentials in the Fast Ionic Conductor Lithium Nitride (Li_3N), *Acta Cryst. A.* **38**, 568–576.

Zurek, E. (2017). The Pressing Role of Theory in Studies of Compressed Matter, *Handbook of Solid State Chemistry* (edited by Dronskowski, R., Kikkawa, S., & Stein, A.), volume 5, p. 571–605, Wiley-VCH, Weinheim, New York.

Acknowledgments

Plenty of figures have been adapted from previously published materials, and I want to thank the copyright holders for granting the permission to include them in this book.

Figures 2.1, 3.1, and 4.4 adapted from Deringer, V. L. & Dronskowski, R. (2013), *Comprehensive Inorganic Chemistry II, Vol. 9*, pp. 59–87, by permission of Elsevier. Figures 2.2, 3.2, 4.1, 4.2, and 4.11 adapted from Dronskowski, R. (2005), *Computational Chemistry of Solid State Materials*, by permission of John Wiley and Sons. Figures 3.3, 3.6, 4.7, and 4.9 adapted from Nelson, R., et al. (2023), *Comprehensive Inorganic Chemistry III, Vol. 3*, pp. 141–201, by dearly paid permission of Elsevier. Figures 3.5, 4.5, 4.6, 4.8, and 4.71 adapted from Müller, P. C., et al. (2021), *J. Phys. Chem. C* **125**, 7959–7970, by permission of the authors. Figure 3.7 adapted from Esser, M. et al. (2015), *Solid State Commun.* **203**, 31–34, by permission of Elsevier. Figures 3.8, 3.12, and 4.29 from Nelson, R. et al. (2020), *J. Comput. Chem.* **41**, 1931–1940, by permission of John Wiley and Sons. Figure 4.10 adapted from Esser, M. et al. (2017), *J. Comput. Chem.* **38**, 620–628, by permission of John Wiley and Sons. Figure 4.13 adapted from Lüdtke, T. et al. (2014), *Inorg. Chem.* **53**, 11691–11698, by permission of the American Chemical Society. Figure 4.14 adapted from Wolff, H. & Dronskowski, R. (2008), *J. Comput. Chem.* **29**, 2260–2267, by permission of John Wiley and Sons. Figure 4.15 adapted from Li, Y. et al. (2015), *Z. Anorg. Allg. Chem.* **641**, 266–269, by permission of John Wiley and Sons. Figure 4.16 adapted from Nakada, A. et al. (2021), *J. Am. Chem. Soc.* **143**, 2491–2499, by permission of the American Chemical Society. Figure 4.18 adapted from Görne, A. L. et al. (2016), *Inorg. Chem.* **55**, 6161–6168, by permission of the American Chemical Society. Figure 4.19 adapted from Missong, R. et al. (2015), *Angew. Chem. Int. Ed.* **54**, 12171–12175, by permission of John Wiley and Sons. Figure 4.20 adapted from Chen, D. et al. (2023), *J. Am. Chem. Soc.* **145**, 6986–6993, by permission of the American Chemical Society. Figure 4.21 adapted from Dolabdjian, K. et al. (2018), *Dalton Trans.* **47**, 13378–13383, by permission of The Royal Society of Chemistry. Figure 4.22 adapted from Corkett, A. J. et al. (2019), *Inorg. Chem.* **58**, 6467–6473, by permission of the American Chemical Society. Figures 4.23 and 4.24 adapted from Bielec, P. et al. (2019), *Angew. Chem. Int. Ed.* **58**, 1432–1436, by permission of John Wiley and Sons. Figures 4.25, 4.41, and 4.42 adapted from Ertural, C. et al. (2019), *RSC Adv.* **9**, 29821–29830, by permission of The Royal Society of Chemistry. Figure 4.26 adapted from Zucker, U. & Schulz, H. (1982), *Acta Crystallogr. A* **38**, 568–576, by generous permission of the International Union of Crystallography. Figures 4.27, 4.28, 4.30, and 4.31 adapted from Ertural, C. et al. (2022), *Chem. Mater.* **34**, 652–668, by permission of the American Chemical Society. Figures 4.32 and 4.33 adapted from Maintz, S. et al. (2013), *J. Comput. Chem.* **34**, 2557–2567, by permission of John Wiley and Sons. Figure 4.34 adapted from Tokmachev, A. M. et al. (2010), *ChemPhysChem* **11**, 384–388, by permission of John Wiley and Sons. Figure 4.35 adapted from Hoepfner, V. et al. (2012), *Cryst. Growth Design* **12**, 1014–1021, by permission of the American Chemical Society. Figure 4.36 adapted from Hoepfner, V. et al. (2012), *J. Phys. Chem. A* **116**, 4551–4559, by permission of the American Chemical Society. Figure 4.37 adapted from

Deringer, V. L. et al. (2014), *Chem. Commun.* **50**, 11547–11549, by permission of The Royal Society of Chemistry. Figure 4.38 adapted from George, J. et al. (2014), *J. Phys. Chem. A* **118**, 3193–3200, by permission of the American Chemical Society. Figure 4.39 adapted from George, J. et al. (2015), *ChemPhysChem* **16**, 728–732, by permission of John Wiley and Sons. Figure 4.40 adapted from George, J. & Dronskowski, R. (2017), *J. Phys. Chem. A* **121**, 1381–1387, by permission of the American Chemical Society. Figure 4.43 adapted from Eickmeier et al. (2020), *Crystals* **10**, 184, by CC BY. Figure 4.45 adapted from Landrum, G. A. & Dronskowski, R. (2000), *Angew. Chem. Int. Ed.* **39**, 1560–1585, by permission of John Wiley and Sons. Figure 4.46 adapted from von Appen, J. et al. (2010), *J. Comput. Chem.* **31**, 2620–2627, by permission of John Wiley and Sons, and from Dierkes, H. et al. (2013), *Z. Naturforsch.* **68b**, 1180–1184, by permission of De Gruyter. Figure 4.47 adapted from von Appen, J. & Dronskowski, R. (2011), *Steel Res. Int.* **82**, 101–107, by permission of John Wiley and Sons. Figure 4.48 adapted from Lintzen, S. et al. (2013), *J. Alloys Compd.* **577**, 370–375, by permission of Elsevier. Figure 4.49 adapted from von Appen. J. et al. (2014), *J. Comput. Chem.* **35**, 2239–2244, by permission of John Wiley and Sons. Figure 4.50 adapted from Timmerscheidt et al. (2017), *Metals* **7**, 264, by CC BY. Figure 4.51 adapted from Clark, W. P. et al. (2017), *Angew. Chem. Int. Ed.* **56**, 7302–7306, by permission of John Wiley and Sons. Figure 4.53 adapted from von Appen. J. et al. (2006), *Angew. Chem. Int. Ed.* **45**, 4365–4368, by permission of John Wiley and Sons. Figure 4.54 adapted from Wessel, M. & Dronskowski, R. (2010), *J. Am. Chem. Soc.* **132**, 2421–2429, by permission of the American Chemical Society. Figure 4.55 adapted from Schneider, S. B. et al. (2013), *J. Am. Chem. Soc.* **135**, 16668–16679, by permission of the American Chemical Society. Figures 4.56 and 4.57 adapted from Luo, D. et al. (2021), *Angew. Chem. Int. Ed.* **60**, 486–492, by permission of the authors. Figure 4.58 adapted from Luo, D. et al. (2022), *J. Am. Chem. Soc.* **144**, 5155–5162, by permission of the American Chemical Society. Figure 4.61 adapted from Benz, S. et al. (2022), *Inorganics* **10**, 132, by CC BY. Figure 4.62 adapted from Decker, A. et al. (2002), *Z. Anorg. Allg. Chem.* **628**, 295–302, by permission of John Wiley and Sons. Figure 4.64 adapted from Deringer, V. L. et al. (2015), *Adv. Funct. Mater.* **25**, 6343–6359, by permission of John Wiley and Sons. Figure 4.65 adapted from Deringer, V. L. et al. (2015), *J. Mater. Chem. C* **3**, 9519–9523, by permission of The Royal Society of Chemistry. Figure 4.66 adapted from Deringer, V. L. et al. (2015), *Adv. Funct. Mater.* **25**, 6343–6359, by permission of John Wiley and Sons, and from Deringer, V. L. (2014), *Quantenchemische Modellierung und Bindungsanalyse komplexer Feststoffe*, Dissertation, RWTH Aachen University, by permission of the author. Figures 4.67 and 4.68 adapted from Küpers, M. et al. (2017), *Angew. Chem. Int. Ed.* **56**, 10204–10208, by permission of John Wiley and Sons. Figure 4.69 adapted from Simons, J. et al. (2021), *RSC Adv.* **11**, 20679–20686, by permission of The Royal Society of Chemistry. Figure 4.70 adapted from Hempelmann, J. et al. (2021), *Adv. Mater.* **33**, 2100163, by permission of the authors. Figures 4.72 and 4.73 adapted from Hempelmann, J. et al. (2022), *Angew. Chem. Int. Ed.* e202115778, by permission of John Wiley and Sons.

Index

https://doi.org/10.1515/9783111167213-009

www.ingramcontent.com/pod-product-compliance
Lightning Source LLC
Chambersburg PA
CBHW081528220326
41598CB00036B/6368